Thinking Pedagogically about Educational Technology Trends

Prioritizing Teaching and Learning Activities with 21 Popular Educational Technologies and Digital Trends from 2016

Jeremy Riel

ETN Press
The Educational Technology Network
Springfield, Oregon, USA

Copyright © 2017 by Jeremy Riel

All rights reserved. No part of this book may be reproduced in any form by any means, print or electronic, including photocopying, recording, or data storage system, without written permission from the publisher.

Some articles in this book were previously published on Jeremy Riel's online Medium publication

Published by ETN Press
The Educational Technology Network
Springfield, Oregon
http://press.edtech.network

Printed in the U.S.A.

Cover design by vicovers

Paperback edition ISBN 978-1-946712-00-4
eBook edition ISBN 978-1-946712-01-1

Dedicated to Jessica and Josephine

Contents

Why Should We Think Pedagogically about Educational Technology? 1

Part 1: Popular Tools and Innovative Ideas

1. Trends in Mobile Chat Apps 5
2. Talking with Computers: How Chatbots Can Help with Learning 13
3. Story Snippets, Ministories, and Microblogging 23
4. Collaboration in Cloud-Based Writing Apps 33
5. Digital Demonstrations Have Become Quite Sophisticated! 39
6. Everyone's Got a Bit of Librarian in Them (Archiving Stuff) 49
7. Pokémon Go and Education 61
8. Making Sense of Minecraft for Learning 71
9. Augmented Reality in the Classroom 79
10. Virtual Reality in the Classroom 88
11. Does This Book Have Pictures? (Using Images) 91
12. Infographics to Infuse Data into Classroom Communications 101

Part 2: Trending Topics and Stuff to Consider

13. What Makes Technology "Smart" in the Classroom, Anyway? 111
14. Thinking about the "Making" Movement for Education 117
15. Learning to Trust the Crowd (User-Submitted and Crowdsourced Info) 125
16. Unintended Outcomes of Technology Use (a Pokémon Go Case Study) 135
17. Mastering the Search Bar 143
18. Toward Criteria for Evaluating the Quality of Digital Information 151
19. Combating Cyberbullying in the Classroom 161
20. BYOD: Bringing Students' Mobile Devices into the Classroom 167
21. Finding Free, Web-Based Teaching Resources 173

Index 179

About This Book

This book represents a year's worth of thoughts and conversations I've had on ed-tech and its associated buzzwords that had been thrown around at the college at which I work. In the chapters that follow, I tackle some of the trending ideas and news on how technology is being used in innovative ways and answer some of the questions that arose in conversations that I had with colleagues in 2016 about trends in ed-tech for both higher education and K-12 classrooms. Drafts of the chapters in the book were previously published as blog posts for students and faculty at the UIC College of Education in order to promote conversation and thinking about *how we use* learning technologies rather than simply putting tech in courses because it's the cool thing to do. I wanted to extend the work that I've been doing at UIC to the broader community, so I have bundled some of my work into this book.

    The essays in this book are designed to be conversational and less-formal so to make the technologies and activities with them approachable for educational practitioners. Every chapter in the book examines a technology in depth with a focus on how educators can directly apply it for learning. You don't have to read the chapters in order, and each chapter is stand-alone in its content. Whenever a specific app, other technology, service, or company is discussed, a citation appears at the bottom of the page to help you quickly look it up and save some time – no need to thumb through the endnotes to find what you're looking for.

    I hope that you find the articles in here helpful for making sense of some of the buzz surrounding ed-tech. If you have any stories on how you have used some of these technologies, or would like to comment on the content of the book, feel free to reach out to me!

Best wishes,
Jeremy Riel
twitter.com/@jeremyriel

Why should we think pedagogically about educational technology?

What does it mean to think pedagogically about technology, anyway? Anytime you use educational technology, it helps to try to think pedagogically about the technology in four ways: learning theory, actions with the tech, goals, and strategies.

In my work as an educational technology consultant and digital pedagogy researcher, I think every day about new technologies for teaching and learning. There are so many cool apps in the news that can help us learn some math skills, keep track of our schedules and to do lists better, or help promote critical thinking. Every time I log on Facebook or check my email, it always seems like there's some new ed-tech app out that I should find more out about. Similarly, after every summer, teachers often find out about a new platform that teachers are required to implement at the start of each new school year.

But let's take a step back - how do we start to prioritize all these technologies for teaching and learning in the first place? Doesn't educational technology already prioritize teaching and learning? Well…not necessarily.

We often hear that training will solve everything, but we know that doesn't always do the trick. Although there are lots of cool things that are designed each year for teachers and classrooms, it seems more often than not that we would really benefit by just taking a step back for a minute and thinking about how these technologies can be used in the first place before throwing them into real teaching and learning situations. That's why I try to think first and foremost about technology **pedagogically** when I work with teachers, students, and college professors in their tech integration projects.

Our main goal for when we think pedagogically about ed-tech is to imagine first the pedagogical implications for technology and how we would want to *teach* with it. We start by asking four questions about the activities and teaching strategies teachers and students related to the technology. Bottom line, it's all about what we end up doing with the technologies in our lives in how they end up influencing learning, not what the technology necessarily does or what the designers say it's for.

So, what are these four questions that we should ask? First, we think about how using this technology to teach and learn would align with what we know about the **theories of learning**. For instance, if I want to use some digital tools in a project-based course, I need to make sure it doesn't promote lecturing or boring memorization drills, which would defeat the theory of project-based learning. We need to figure out whether or not the use of a particular tech aligns with our commitments to how we want to teach. This prevents us from allowing the tech to drive us toward undesired outcomes and keeps us in the driver's seat. Second, we should ask questions about what types of **actions the technology** allows teachers or students to do (in what we sometimes call *affordances* in educational research). Third, we should look into the different **goals** that apply to teaching and learning. Finally, we need to consider the **teaching strategies** that can be used with the technology, and not just what the students can do with the technology: that's the *pedagogy* part of the equation. When we consider what we can do, our goals, and the teaching strategies we can employ, we get a *principled digital pedagogy* with potentially transformative effects.

When I think about ed-tech tools, I avoid focusing on individual technologies or platforms, such as exclusively examining only Blackboard, Google Docs, or Edmodo for learning. Instead, I look at bundles of similar types of technologies to think about how we could use them: *what do we want to do with the tech?* This is an approach that allows us to be better adopters of individual educational technologies, even if we've never encountered a specific technology before. Focusing on the pedagogy, or the actions of teaching and learning, allows us to see past individual platforms or tools and focus on our objectives as educators. Otherwise, it becomes too easy to just throw expensive technology at a problem and see few results. Technology is only as good as how we use it.

These are the four questions that I ask when I think pedagogically about a technology. There's no right way to do it, and I've found that it really is something that just gets better with practice. You don't have to write a report or give a presentation on your findings, but sometimes some quick research can pay dividends. Try tossing these questions around in your head when you're thinking about new tech for teaching and learning to get a better idea of how it might be used in authentic learning situations or how it could influence learning.

Part 1

Popular Tools and Innovative Ideas

Trends in Mobile Chat Apps

1

Making the most of chat for learning while not ruining the fun

If you were to ask yourself what the under-25 crowd does most on a mobile device, you'd probably conclude that chat, text, and instant messaging apps draw a good portion of attention. Instead of just the basic text messaging that dominated mobile devices for years, students today carefully curate and send information on many platforms to keep in touch with friends and track the trends and news that they care about. The influx of messaging apps and new functions available to chatters are of interest to educators, as these everyday apps for students can fundamentally influence the ways in which students communicate and process information. Students use chat apps every day for numerous purposes, so using these apps is an alluring hook for educators to get kids engaged and interested in learning activities.

It's a classic scene today: Kids all in the same room not saying a word out loud, but chatting with each other in real time via their digital devices. Sometimes you'll hear a muttering of a sound or a giggle, but K-12 students today are very comfortable with having active digital conversations with people in close physical proximity. History has shown that language has continually evolved over long periods of time. However, the means and media of communication today are changing more rapidly than the language itself. This creates entirely new mashups of language and meaning that can be hard to keep track of. As educators, it's important to keep an eye on the changing trends in how youth communicate so that we can continue to reach students in personalized ways.

Chat apps being used today: The usual suspects

On a typical under-25-year-old's phone, you can find any number of apps being used for messaging and communication. There are many of these apps in the app stores, far too many to list here. Here are the big players right now in chat apps for K-12 and college students (with this list becoming obsolete in less than a year, I'm sure!):

Snapchat[1]. An esoteric app for most adults, Snapchat is an app in which participants send messages back and forth with little planning or consequence, as messages only are visible for a few seconds. It has an easy learning curve, but expects users to share media-rich communications…that is to say that text is usually not enough for Snapchatters. As such, most "snaps" are a mélange of images, videos, texts, and annotations that, combined, uniquely communicate ideas. It promotes the ephemerality of information, much like how an in-real-life conversation would go. It's probably the most popular messaging app of the under-25 crowd.

Kik[2]. An app that's popular with the younger crowd, it's a full multimedia chat app that can run over Wi-Fi. One thing that has gotten Kik in the news lately is its interests in pursuing chat bots. These

---

[1] App- Snapchat: http://www.snapchat.com
[2] App- Kik: http://www.kik.com

are digital robots with which users can interact and have more personalized experiences. Chat bots in a messaging app can be used for games, information, or other help from companies, friends, or online communities.

WhatsApp[3]. One of the original phone messaging apps over Wi-Fi, this app allows for multimedia communications using video, camera photos, audio, emoji, stickers, images, and text. It may have been the app that got people excited about the new frontier of messaging in the first place, as it allowed people to express themselves in ways that boring old texts couldn't do. It now has millions of users and supports group chats, as well as all of its multimedia functions.

SMS: regular ol' text messages. Yes, despite all the hype about and use of messaging apps, SMS, or the "simple messaging service" that you use to message phone number-to-phone number is still used by a lot of youth, but only IF they have a wireless data plan. The promise of chat apps is that you can use them on Wi-Fi networks, which are more common today than they were when SMS text messaging became prominent.

"Mainstream" messaging apps: Facebook Messenger[4], **Google Hangouts**[5], **Google Allo**[6], **Skype**[7], **and Apple Messages**[8]. These apps continue to be used by the under-25 crowd, but not at the same level of use as people in older cohorts. Again, these apps will be used only if other friends, colleagues, or online communities (e.g, gaming communities) use the apps. They're the popular, more mainstream apps, but adoption depends on where a user's social network interacts.

---

[3] App- Whatsapp: https://www.whatsapp.com/
[4] App- Facebook Messenger: https://www.messenger.com/
[5] App- Google Hangouts: https://hangouts.google.com/
[6] App- Google Allo: https://allo.google.com
[7] App- Skype: https://www.skype.com/en/
[8] App- Apple Messages: https://support.apple.com/explore/messages

Features of chat apps:
Why are they used so frequently?

On their surface, they're just simple messaging apps. At their most basic of functions, a user sends text messages to another person or group of people. However, modern messaging apps have some new functions that have found their way into the technology. These have fundamentally altered the ways in which people communicate via the apps.

Video, images, and a full media experience. We don't communicate in text only, although we have hundreds of years of history to contend with in this regard. Most surviving media over the ages are text-based, but in recent years, images, audio and, video have found their way into our communications. Chat apps have extended past text-only to now allow many forms of expression, each of which carry unique meanings. To convey more complex ideas, modern chat apps allow for media to be merged and remixed to create unique messages to express one's ideas.

Different media, different meanings. Based on the theories of digital and multiple literacies, spoken words may hold different meanings than their digital chat counterparts, and these meanings are only good in certain contexts. A message that may have one meaning in school may have an entirely different meaning (or non-meaning) within a group of friends. Students today that master chat apps are becoming masters of codeswitching and navigating different meanings implied by digital app communications.

Media choice. More media choices open the doors for using analogy and metaphor, or references to other cultural objects that have meaning (e.g., internet memes[9]). It's important for educators to pay attention to these unique combinations of digital media use in order to understand how students think and communicate. For some people, an annotated photo of where they're at on Snapchat speaks volumes more than words alone, or a meme or emoji might convey meaning much better than a sentence.

Ephemerality. Snapchat calls it "living in the moment," which has some merit if you don't have to worry about a chat history or

---

[9] Wikipedia (n.d.). Internet Meme. Web article.
https://en.wikipedia.org/wiki/Internet_meme

that you should polish your prose before sending (unlike an email...). I say ephemerality in the sense that you don't have to worry about how your words may be taken later on down the road, because there likely won't be any record of it. When it comes to photos, right-now and never again images are appealing because you don't have to worry about how bad a photo might look, or worse, how bad your selfie might look. It won't be shared for more than 10 seconds. The "spur of the moment" and unrecorded aspect of some of these apps also give us a chance as a society to look back on how we've been using technology and to question what should be and shouldn't be put into the permanent record.

Multi-platform. I mean this in the sense that the same account can be accessed on any device in the cloud regardless of the make. In many cases, multi-platform apps also means you can chat in an internet browser or on a desktop/laptop computer, giving users more access to their communications tools. In comparison, Apple Messages only works on Apple devices. The multi-platform aspect of chat apps is quite valuable as devices become more diverse.

Thinking pedagogically about chat apps

There are a few things to ponder when considering the application and impact that chat apps can have on learning activities. Consider some of the following points when planning a classroom activity in which you would like to try to use some of the communications tools that are being used by students every day.

Access to messaging tools. Some digital tools are still not ubiquitous among all students of certain ages, even it seems every kid has a phone in their hands. Not all students have digital devices, or, more importantly, devices with wireless network plans attached. Some students' devices will only work on Wi-Fi, limiting them to the amount of time they can work. Be aware of your students' access to the base technologies you employ before assigning work. One of the features of modern chat apps is that they work seamlessly across multiple devices and allow computer-based desktop access, which can improve students' access to apps used in the classroom.

Influence on group work. Even if you don't encourage the use of chat apps in class activities, these apps will likely influence group work, especially out of the classroom. It's just a matter of fact now that these apps are used by many students to communicate with each other. It really gets us thinking about the "proper" and "best" way for students to communicate. In a true philosophy fashion, i'm iffy on requiring students to only use the technologies in the business world, as it is the youth that typically innovate and shift our society's ways of doing things. If I had to take a stand, I'd say that it should be balanced and that students should be able to communicate in various circles, including their own. It may also be the case that students may not want school in their digital communications. These apps are likely avenues to tap into how students digitally communicate in their everyday lives. However, it is wise to test the waters and ask students about the apps they like to use to avoid appearing clueless to students. It's important to use apps with a purpose in mind and not just because "it's cool with the kids."

On-demand help and differentiation. By communicating with students on their terms, they may be more apt to asking for help and communicating their needs. By setting some terms for communicating with teachers, educators might unlock some great potential in reaching students' specific needs and differentiate instruction accordingly. It may take a bit more work to get used to a new chat app, but personalized help might be more effective if educators use the communications tools with which students are comfortable and want to use. Find out what chat apps your students are using and pick one or two to accept messages from students. Just remember to set some guidelines on when responses should be expected and what kinds of behavior are appropriate if chat apps are integrated into class work.

Multimodal literacy development. Video, audio, text—and the combination of these—are all key components of chatting on today's messenger apps. Each medium requires separate literacy skills to understand the meaning behind messages, and meanings change based on the audience and contexts. Critical literacy skills could be taught and promoted by use of multimedia messaging apps. Getting familiar with these different ways of communicating might be useful for teachers to understand how their students communicate. It may also be helpful for developing strategies for developing other communication skills in your students, such as creative writing, long-form writing, communication through more business-y channels like email, and in-real-life speech.

Reminders and notifications on platforms students use can help students stay organized and help teachers communicate regularly with students. Some reminders can even be automated with tools such as IFTTT or by setting up a simple chatbot (see Chapter 2).

Considering challenges with chat apps. There are many challenges that go along with embedding communications technologies in class. First, using multimedia and rapid-fire communications technologies helps students grapple with the balance of their real and digital lives. It's a good opportunity to discuss the trade-offs we get in life by being always connected. Using too many chat apps, or getting messages too frequently may lead to anxiety, distraction, and not knowing how to handle quiet, disconnected time. Second, bullying from classmates or internet strangers remains an issue when using digital communications tools. I've discussed this more in depth recently in an article that readers might find helpful. Finally, students have opportunities to consider the challenges with data permanency in today's app culture. Some apps promote the immediate deletion of communications, while some will store data indefinitely. When using tools like chat apps, it's a good opportunity to discuss how permanent data storage and distribution can affect one's life, and how data can be easily duplicated in today's apps even when you think it's private (such as when people take screenshots of seemingly impermanent Snapchats).

Talking with Computers: How Chatbots Can Help with Learning

2

The recent rise of chatbots and their educational uses

Followers of tech trends may have noticed the word "chatbot" appearing in an increasing number of news publications over the last few months. **Chatbots,** or automated software for chatting with a computer, have been around since at least the 1960s. With the recent rise of popularity in chat apps among youth, however, they have reached a newfound fad status in the last year or so. But what is it about these talkative software-based robots that has people excited and companies and organizations scrambling to set up their own chatbots? More importantly, what potential do they have for learning? In this chapter, I tackle what defines a chatbot, what they can help people do, and how educators can think pedagogically about this old technology that's gotten a lot of new interest.

Chatbots have found their way into discussions around Silicon Valley and educators' circles alike. Advances in ubiquitous chat apps among teenagers have prompted a new way of thinking about how people interface with information and organizations. This has encouraged tech developers to bring about a whole new generation

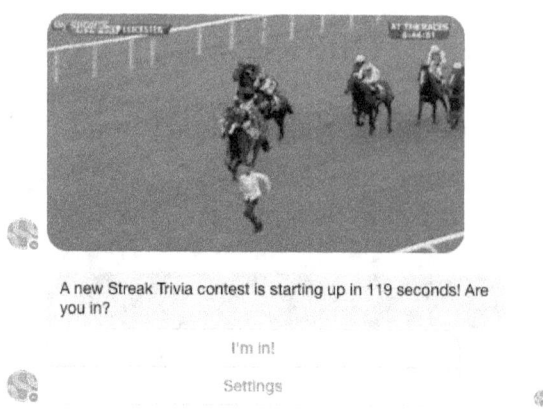

*A new game of Streak Trivia is beginning.
Are you in or out? You only have a couple minutes to respond!
(a screenshot of a Streak Trivia chatbot game that I played)*

of talking robots that you can chat with as if they are on your friends list. Businesses and organizations are using it, opening chat windows for customer service, help desks, FAQs, and for directly placing orders for things. Need a shirt of a certain color or a new pair of shoes? Just send a message to a chatbot at Nordstrom or Macy's and they'll help you right out! You can have CNN's chatbot send you news[1] whenever something interesting happens in the world (which, in the age of every-hour news, your feed could be flooded!). Want to know when you need an umbrella? There's a chatbot for that[2]. The White House even recently set up a chatbot to help field questions from the public[3], although the jury's still out on how well their experiment met expectations.

---

[1] App- CNN Chatbot: https://www.messenger.com/t/CNN/
[2] App- Hi Poncho: https://www.messenger.com/t/hiponcho/
[3] Peterson, A. (2016). I tried the new White House Facebook chatbot. Here's what happened. *The Washington Post.* 10 August, 2016.
https://www.washingtonpost.com/news/the-switch/wp/2016/08/10/i-tried-the-new-white-house-facebook-chatbot-heres-what-happened/

Photo credit: Steve Rainwater (via Flickr-CC)

You can even play games with chatbots. I recently added a chatbot called Streak Trivia[4] to my Facebook Messenger friends list. This bot hosts a daily true/false trivia game for anyone who wants to play. It's simple: the bot will ask you many true/false trivia questions. The person who has the longest streak of correct answers wins.

A brief, long history of chatty software

There are a lot of new chatbots that have arrived on the scene in recent months, with new ones being created every week. But why are they so popular again, considering this is a technology that has been around since the 1960s?

Simply put, a chatbot is a robot that chats, usually in text messages. Chatbots don't exist in the physical world unlike their more fleshed out cousins. Instead, they are software that offer pre-programmed or computer-generated responses to whatever users put it. They also tend to come with a lot of limitations, in that they

---

[4] App- Streak Trivia Chatbot: https://www.facebook.com/streaktriviabot/

can only do what they were designed to do. It's a digital chat buddy that can only respond to a certain range of questions.

Chatbots have likely gained popularity again today because dialogue is a more natural interface for interacting with the world, and chat apps have gained a significant foothold in the daily interactions of most people with handheld digital devices. We, as humans, communicate through conversation on an almost daily basis. Instead of the clicks of a mouse, and even the touch of a tablet, a more intuitive interface may be the text and speech we use to interact with other people.

Chatbots aren't anything new, but they may be enjoying the right conditions today in how people use technology to communicate for them to gain popularity and attention from developers. We're seeing a resurgence in automated chatting due to the increased popularity in chat apps by the under-25 age group in recent years, toward perhaps a more intuitive interface for interaction: dialogue. It's a buzzword that we're starting to see pop up in education again as well, with educators and policy makers wondering how we can leverage these chatty robots to improve learning. But the history of chatbots show that it's not an easy tech *to get right*. It's tough to create a robot that can talk like a human.

Although this seems promising, it is a tough tool to make well. Since the 1960s, computer scientists have been working on ways to have computers understand and work with language in ways that make sense (a field called *natural language processing*). It is wonderful that chatbots are back in the news, as it represents some advancements in the field and a focus on their potential. However, they're not so simple to put together, especially for education. As technology developers renew their interest in chatbots, it will be important for educators to critically examine the offerings that chatbots provide and to evaluate how chatbots can be used toward pedagogical goals.

One of my earliest experiences with a chat-like program was in the text-based video game **Adventure**[5], which incidentally is a great illustrative example for someone wanting to understand chatbots better. The game was written in the 70s, but I got my hands on it with my old DOS computer in the 80s. As a kid, I played the role of

---

[5] Wikipedia. (n.d.). Colossal Cave Adventure. Web Article.
https://en.wikipedia.org/wiki/Colossal_Cave_Adventure

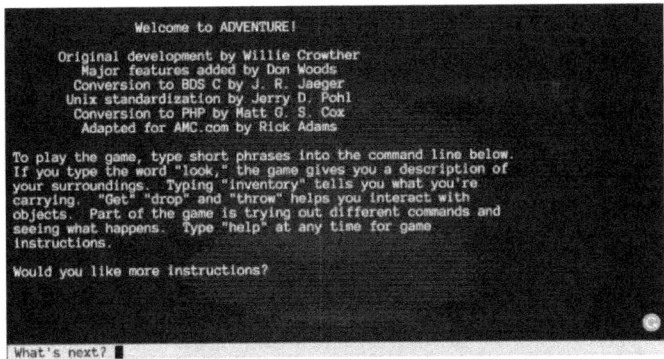

A screenshot of Adventure. All commands in Adventure have to be input with text. The computer is looking for specific words and phrases, very similar to how many chatbots today function.

the adventurer who was exploring a dark cave. All actions of the game had to be input as text, and you received text in response from the game. There were no pictures - only words. Everything the system output was in response to what you put in as the player. If the system didn't recognize your words, you had to repeat yourself in a different way or find a new set of keywords to give the system to progress in the game.

You can play a working version of Adventure right in your browser[6]. Give it a try and see how far you can get!

Over the years, things started to get better for chatbots. Developers of chatbots tried to mimic human conversation better and programming got more sophisticated. The developer community is a pretty robust group, with the community competing annually for the coveted Loebner Prize[7] for most human-like chatbot (have a conversation with "Mitsuku"[8] to see a cool, functioning example of

---

[6] App- Adventure Game (via AMC): http://www.amc.com/shows/halt-and-catch-fire/colossal-cave-adventure/landing

[7] Service- Loebner Prize: http://www.loebner.net/Prizef/loebner-prize.html

[8] App- Mitsuku: http://www.mitsuku.com/

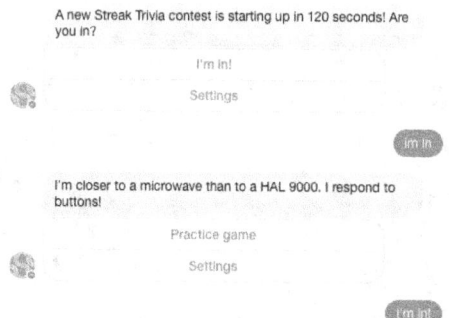

Streak Trivia's identity as a microwave and not HAL 9000 is a subtle reminder that despite the advances in chatbots, we have nothing yet to worry about with chatbots taking over the world.

a recent Loebner winner). Each year, new approaches are invented and previous approaches are refined. Today, machine learning principles are increasingly used in advanced chatbots to allow the system to learn how humans interact with it, giving it better responses. Regardless of the approach, though, the premise remains the same: a user will ask the chatbot a question, or make a comment, and the chatbot will reply with a suitable response. If it didn't understand, it will say it didn't understand and ask the user to repeat or try saying their question in a different way.

Many modern chatbots still don't know much about language, though. In a recent game of the Streak Trivia chatbot I mentioned above, I tried to input my commands in a way it didn't understand (because my typing wasn't exact). It prodded me to communicate with it in a more productive way with a touch of humor.

Thinking pedagogically about chatbots

Despite the constraints of chatbots, the education world quickly saw benefit in early chatbot technologies of the '60s and '70s. Educational technologists invented approaches like **cognitive tutors**[9] and "intelligent" interaction systems in the early days of AI and chatbots to work with learners without needing the intensive time

---

[9] Wikipedia. (n.d.). Cognitive Tutors. Web Article
https://en.wikipedia.org/wiki/Cognitive_tutor

required for one-on-one coaching. Although these systems have a loyal following of researchers, the recent attention and buzzword status of chatbots might bring much needed experiments and investment that the field needs to realize any educational potential. The newest generation of chatbots have some capabilities that can reach students in ways that may not have been achievable with past technologies.

Personalization. The main selling point of chatbots for education is the personalization that one can get. You can have a conversation that is tailored to your specific questions. However, this really only works if we know students' unique needs and communication styles. The potential for adaptation is huge, as has been seen in ed-tech groups like **Khan Academy**[10] and **Knewton**[11]. But, as chatbots get made for education, it is important to make sure chatbots align with the same communications tools that students already use, or else they risk being put to the side of more favored apps. We should also make sure that the adaptive principles that are used in these technologies align with modern learning theories. Educators also need to make sure that adaptive chatbots for learning avoid lumping students into categories or clusters that may disadvantage them later on by preventing future learning, such as any kind of profiling a student as "low performing," which could then lead to chatbots delivering students content that is geared toward this bias in the design. To maintain a commitment to equity in education, adaptive chatbots should maintain an openness in their design that allows students to grow and progress, and that avoids any kind of response pattern from the chatbot based on their performance or use of the chatbot.

Monitoring and nudging. Although we lack good chatbots that are intended for education, some chatbots out there now show some of the possibilities for chatbots for learning. Even though a user might know that they are interacting with machines, recent chatbots have been shown to be effective monitors of behavior and can even nudge users in positive directions. By using language, users of chatbots can be reminded of goals and tasks in a more natural way than phone notifications or emails, and can be prompted to reflect on everyday life through the seemingly basic conversation

---

[10] App- Khan Academy: https://www.khanacademy.org/
[11] App- Knewton: https://www.knewton.com/

with the chatbot. A couple of good examples of this have grown in popularity recently. A chatbot called **Fitmeal**[12] tracks meals and food intake and has conversations with users to improve physical health. Another chatbot called **Joy**[13] interacts with people to help improve mental health. Joy chats with users daily to help monitor how you feel, and works with you to help you make decisions that are beneficial to your mental well-being.

A more human vibe. Even if we know it is a machine, a more natural dialogue style can foster personality when interacting with a machine. I'm just hypothesizing here as more studies in this area are certainly needed, but dialogue with machines in human ways can make the whole experience more pleasurable. Instead of the impersonal interaction one has with library catalogs, book indexes, and even course syllabi and frequently asked question lists, a personal, conversant machine may be more effective at communicating information and helping learners. It may also foster students' personal inquiry by prompting students to ask questions and encourage engagement by using one of our species' most original interfaces: interactive language.

Instant help. Functionally, chatbots might be most helpful at providing information on demand, as they are always ready to talk. The robot is always at the other end of the line. Recently, a great example of this was demonstrated when a Georgia Tech professor "hired" Jill Watson[14] (a chatbot) as his teaching assistant. Throughout the whole semester, students didn't know their TA was a chatbot, but found her quite helpful nonetheless. She fielded hundreds of questions and responded with useful feedback. In fact, many students fell just short of nominating her for a TA teaching award before finding out her digital identity. Asking questions and getting help from a chatbot can be beneficial in other ways too, as some students may have anxiety about asking questions to teachers directly. In addition, instructors can review the chat history and sift through the more meaningful questions that the virtual TA can't

---

[12] App- FitMeal: http://www.fitmeal.com/
[13] App- HelloJoy: http://www.hellojoy.ai/
[14] McFarland, M. (2016). What happened when a professor built a chatbot to be his teaching assistant. *The Washington Post.* 11 May, 2016.
https://www.washingtonpost.com/news/innovations/wp/2016/05/11/this-professor-stunned-his-students-when-he-revealed-the-secret-identity-of-his-teaching-assistant

answer and address these questions with students. As such, it can be a source of questions for teachers to answer, sometimes to questions that they may not have ever received in class.

Next steps for chatbots and education

Unfortunately, there aren't a lot of chatbots for specific educational purposes to try yet. Despite all of the hype, educators are still left desiring rich interactive chat experiences that actually work.

One reason for the dearth of educational chatbots is that programming a chatbot is tough. The most advanced chatbots today are specialized around certain purposes with complicated approaches like machine learning, and significant investment in development talent and content is made. We can see this in some of the more business and organizational chatbot applications. However, for the technology layperson educator, there are some tools out there to get started if you'd really like to. For this, I recommend **Chat Fuel**[15] to create a chatbot—no coding necessary! It may not do everything you want it to, but it would be a good first step to testing the chatbot waters. The **Facebook Developer**[16] resources page also offers a bunch of resources for those who would like to set up a chat bot in its Messenger chat app, but require significant coding.

So, it appears educators must continue to play a waiting game for developers to find useful applications for chatbots for learning. It will also be important to maintain strong commitments to learning theories and design principles that are known to foster learning instead of designing experiences that are not beneficial, like pure rote memorization and drilling activities. These apps have their value in certain circumstances, but rote memorization apps and games have a long, ineffectual history that continues to plague the field of educational technologies. Like most technology in education, the field would be best served if educators themselves took to programming the apps we'd like to use. If you have an idea, give it a try!

---

[15] App- ChatFuel: https://chatfuel.com/
[16] Service- Facebook Bots for Messenger:
https://developers.facebook.com/blog/post/2016/04/12/bots-for-messenger/

3

Story Snippets, Ministories, and Microblogging

New pedagogical tools for leveraging short-form digital storytelling

As new media continue to find their way onto the digital landscape, storytellers are finding themselves with new creative ways to share their narratives. The form and content of stories have recently been transformed by the emerging super-short format of what I've been calling **short-snippet storytelling**. These bite-sized chunks of strung-together media have become incredibly popular on the web in the last two years. Although the format is new and can take some getting used to, it has some specific benefits that can be leveraged for learning!

There are multiple reasons educators should start paying attention to the seemingly disjointed strings of posts that I call short-snippet storytelling, which are also commonly known as or **short-form storytelling** or **microblogging**. First, it's popular...very popular. Look at the phone of any under-20-year-old and you'll see them stringing together a story from a bunch of photos, videos, and

24 Story-Snippets, Ministories, and Microblogging

Can we say more with fewer words? Short-form stories are starting to gain in popularity – and are built from only small chunks of media
Photo credit: Speech bubbles at Erg (via Flickr-CC[1])

texts they make throughout their day. Keeping the pulse on tech use by students and mining the pedagogical nuggets are what each of these chapters are all about!

Second, short-snippet digital stories are not just about telling every detail of one's life - those stories will get ignored amid the fast, daily stream of data. Instead, choosing and stringing together information to share to tell a story in creative ways has seemed to grab the attention and eyeballs of the mobile device generation. In short, it's more about people doing interesting things in their lives and telling interesting stories. Figuring out what is interesting, though, is just a matter of knowing and listening to your audience. If it's just a simple listing of one's everyday activities, it may not have much value. There's something about what makes these stories *interesting*, and identifying that element can have deep pedagogical impact.

Finally, because stories are made of many small-effort and short-shelf-life bits, there's a really low risk to participating and thus, a much lower bar for entry. Participants don't have to

---

[1] Wathieu, M. (2010). Speech bubbles at Erg. Flickr photo. http://www.flickr.com/photos/88133570@N00/5263647030

overthink their smaller contributions too much, so it may be easier than ever to get stories made and shared!

Telling a story in 100 chapters: Bite-sized chunks

With new media short stories, creators share very small chunks of a story at a time to a broad, open audience. Stories can be pre-conceived and edited, or they can be told live, in the very moment they are happening. However, instead of stand-alone posts that tell a whole story, the short blurbs are meant to be strung together or seen in a sequence. Think along the lines of Twitter, with many small 140-character snippets, but having some way to read the tweets in order. In fact, Twitter is one of the original forms of "microblogging," as storytellers would share their stories with the world with a series of small text messages, but with the intent that they be read in a specific sequence.

I say that you "read" short-snippet digital stories, but *reading* may not be the right word. With short snippet storytelling in 2016, any media can be used, and the combinations of various media can have interesting effects. For example, on a platform generally known for a text-only interface like **Twitter**, you can now share video (e.g., **YouTube**, **Instagram**[2] posts), images (e.g., **GIFs**[3], **memes**[4]), annotations (hand drawn notes o/tn images and videos), and links to websites in addition to the classic 140-character text that made Twitter famous. The type of media, the timing and sequence of posts, and the commentary by the storyteller all create super rich narratives that help "readers" experience happenings in meaningful ways. Top social networks like Facebook, Snapchat and Instagram are all adopting storytelling tools that allow the use of various forms of media. This leads me to believe that all social media platforms should also include useful storytelling media. A general rule of thumb on social networks is that followers want to see interesting happenings and perspectives of those that they follow.

---

[2] App- Instagram: http://instagram.com
[3] App- Giphy: http://giphy.com/
[4] Wikipedia. (n.d.). Meme. Web article. https://en.wikipedia.org/wiki/Meme

As such, short-snippet stories served up on social networks make it easier than ever to craft a story and share it.

The functions of short-snippet storytelling are a bit different than conventional story publication. Although all stories are built from small elements of events, people, concepts, place, and time, this method prioritizes small elements literally. Instead of using a single, dense text to string together events, participants, and a plot, storytellers find various meaningful elements and share them over time. Storytellers are afforded the opportunity to create brief updates to their stories. As a result, it's a low risk that one specific element will ruin a story, reducing the requirement to think too hard about editing and revision. These smaller chunks are easier to make, easier to digest, and easier to respond to. Media can be produced faster and with lower barriers to entry.

It's worth noting here that I think there are two types of stories shared on these kind of media right now. However, it's hard to specify where one begins and the other ends, so I'll probably dive in deeper on the differences between the two in a future essay. First, there are thought-out and edited short-snippet works, like conventional stories in which there's room for revision and editing. A scripted story, either fiction or nonfiction, can be developed and shared over time on these media. Second, and something we haven't seen in previous media, is live, by-the-moment short-segment storytelling. Using the range of media in the short-snippet form, storytellers can share the stories of their lived experience *as it happens*.

Thus, stories can also unfold in real time. Similar to how journalists tell the news, any person can now share the story of their experience as it unfolds. As such, the upcoming sequence of events is not known to anyone, including the storyteller. The popularity of recent inventions like **Facebook Live**[5], **YouTube Live**[6], and **Snapchat Stories**[7] are all indicators that people are finding value and interest in how others live and see the world without any editorial interference. Although the storyteller chooses the type of media with which to share the story, the *raw, unedited story* is appealing. Readers can follow a story as it unfolds, which can be just as excit-

---

[5] App- Facebook Live: https://live.fb.com/
[6] App- YouTube Live: https://www.youtube.com/channel/UC4R8DWoMoI7CAwX8_LjQHig
[7] App- Snapchat Stories: https://support.snapchat.com/a/view-stories

ing for the reader as it is for the person telling the story! It's like a sitcom back in the day for which you have to excitedly wait each week for new episodes...except only the updates now come faster and are in smaller chunks than a 30-minute episode on primetime.

How do you string a short-snippet story together?

As creators string together the nuggets of a short-snippet story, they can create a chain of tens or even hundreds of posts that tell a story. Many different forms of media come together to paint a vivid picture that the storyteller wants to get across. Unique combinations of media can be used to convey the events, people, concepts, and emotions that the storyteller wants to share. In effect, the storyteller also serves as a curator of media, bringing together the many chapters of the story and serving them up in small servings.

There are many apps and platforms that are being used right now to tell stories. The classic example is how many people use Twitter to tell stories. Instead of stand-alone tweets about something, many twitterers will share a string of posts around a common theme using a variety of methods. On Twitter, it is most common to see a story told over time in a sequence of ordered posts. Additionally, tweets can include a common #hashtag, which are used to easily search any post that uses the unique keyword contained in the hashtag. Posts are also not limited to text: videos, images, and audio can also be shared. In addition to spacing out posts sequentially, live events are also often captured on Twitter in what is called *liveblogging*. When they liveblog, storytellers rapidly post updates about an event as it unfolds. This method allows readers to experience events through the narrator's tweets in real time, much like a live newscast at an event. Finally, Twitter has recently released a tool called **Moments**[8] that allows users to string together a series of tweets to tell a bigger story and distribute it to viewers. For many, it's more about the sequence of posts more than what any individual post contains.

---

[8] Muthukumar, M. (2015). Moments, the best of Twitter in an instant. Blog post. Twitter.com Blog. https://blog.twitter.com/2015/moments-the-best-of-twitter-in-an-instant-0

Like Twitter's strings of stories, YouTube users have been sharing short stories for years. Many YouTubers have gained popularity as *vloggers*, or video loggers, as they share their everyday experiences in the form of short, unedited videos as things occur. Vloggers tell stories as they occur, with their audiences anxiously awaiting new posts. To this end, YouTube has recently allowed users to broadcast their videos live, allowing active audience participation through comments and video responses. In effect, the story can be influenced by the audience as it happens!

Although remaining an esoteric app to educators, Snapchat's Stories feature remains popular for many of the platform's users. Creators can deliver snippets of video, photos, and annotation over time and string them together in a specific order. Brands (through **Snapchat Discover**[9]), organizations, and celebrities create sequences of media for their followers, giving them a stream of media doses with each providing a deeper meaning than any one single snippet can provide. In fact, Snapchat even briefly flirted with the idea of providing original programming[10] on their Stories platform, much like a Netflix of social media. However, the most important technological contribution of Snapchat Stories is that stories only last 24 hours. This gives rise to the idea of a *digital shelf life*, something that hasn't been considered in the age where the admonition that "Google Never Forgets" rings too true. The function must be popular though, as Instagram recently implemented a similar function to tell strung-together stories on their social network, and Facebook has indicated an interest in the same.

As we can see, this **shortest**-form storytelling method has provided some unique tools to storytellers that have not been available in the past. For the first time, small chunks can be distributed widely to followers. Stories can have permanent homes on the web, or they can have daily sell-by dates. As such, time has become as much a tool for storytelling as the myriad media that narrators have at their disposal. The stream of snippets can be controlled and curated in ways that afford great opportunities for experimentation, audience engagement, and easy entry to creation.

---

[9] Service- Snapchat Support: https://support.snapchat.com/en-US/a/discover-how-to

[10] http://money.cnn.com/2015/10/13/technology/snapchat-abandons-original-programming/

Thinking pedagogically about short-snippet stories

There are volumes of research on the benefits of getting students to tell stories and the use of narratives for learning, so I won't get into that work here. What I will discuss, however, is that these micro-media stories help better emphasize the events in creative work and make it easier to create as events occur. The platforms for producing and sharing short-snippet stories promote the authenticity of stories as a way of expression, making sense of the world, and sharing with others.

Shorter might be better at some things. When "reading" a short-snippet story, these media can be digested over time, and in smaller amounts. It requires far less investment than a book or even a blog post, which could possibly improve motivation and interest. Stories are sent directly to users' inboxes on these platforms, so it's easy to stay up to date on a story's progress. In short, it keeps viewers anticipating more...as long as the story stays interesting, that is.

Smaller bits of media also mean that many smaller, measurable achievements get done. It's much easier to track one's work and progress with smaller chunks of work. The size of a snap, tweet, vine, or short YouTube video is manageable and well-defined in this realm of storytelling. Having a bunch of smaller tasks may be easier to complete and make larger projects look less daunting, much like how modern video games use small quests, missions, and small achievements to promote progress in the game.

These media are redefining the units of daily communication. to the tweet or vine instead of the sentence, paragraph, page. This may not necessarily be a bad thing, as smaller chunks are easier to interact with. However, they can be hard to take as a whole, and could lead to many tangential distractions while trying to accomplish work.

Multimedia affordances: things "just text" may not be able to do. Small-topic stories told with the short-snippet mode can use video, audio, images, and yes, text. For many K-12-aged students, this beats just having to write solely with words. Although text literacy is an important skill, some media might convey ideas in a better way than text can provide. What's new about the short-snippet story method is how it *ties together* unique combinations of media.

Storytelling platforms like Snapchat and Twitter specialize in tying these media together into chains, giving rise to the unique combinations of communications tools. In addition, each of these media can have multiple layers of annotation, with storytellers able to draw on, add text, and provide additional information on the surface of images and videos. Storytellers can also link to outside resources to support the story or provide context. In today's digital landscape, teachers who use these tools in the classroom can promote the development of digital literacies that empower students to create and understand meaning with digital media.

Low barrier to participation brings increased engagement. With these chunks of media being so short, there's a much smaller learning curve for both creators and consumers. You can jump in much easier and not have to worry about your early content not having high production value. What's important is the ongoing story, not any individual post. The tool also gives voice to students who may not typically speak up in class. With the low risk of failure due to the lack of longevity in short posts, anyone can approach these media with their own interests and communication styles to find their voice through a unique combination of storytelling tools. As a result, teachers can be a member of the audience for students' stories to see how students see the world and think.

Two-way street: Storytelling is not just for the writer. These media also allow the audience to interact with the author. With smaller chunks, it's easy to jump in and catch a story as it is being told. Given the medium, it's less important to know the backstory and to catch up than it is to follow along as stories play out. In addition, the common platforms like Snapchat, Twitter, Instagram, and YouTube promote audience interaction. The authors of stories can answer questions, respond to feedback, and even allow the audience to influence how a story continues to get told.

Some additional thoughts

Despite all of the benefits I identified above, I'm not yet sure if short-snippet storytelling is a good thing for learning. Most importantly, its disjointed and hyperlinked form may reduce focus (something of which I'm definitely guilty!) For instance, in contrast to a string of tweets and short videos, a book bundles everything together and isn't so all over the place, which can improve focus. Even a Wikipedia article or news article bundles a story together, making it easier to read. Will strung-together stories exacerbate our already addled attention? I don't think I can venture a guess yet to answer that question.

However, in the modern media age, the idea of *a stand-alone media document* is becoming a rarity. This book is a good example as it is full of URLs and references to other documents. I encourage readers to check out sites on their phones while they're reading the book and try out some of the resources I suggest. Even the examples I cited above aren't that great at keeping attention in one place. News stories today typically have additional links to previous news stories sprinkled throughout the article. The defining characteristic of Wikipedia pages are their hyperlinks scattered throughout a page, linking readers to hundreds of other pages. As such, much of today's media encourage tangents in people's media consumption. Focus is not a new concern in media use, so there's much research to be done on this method before any speculation can be asserted about it being good or bad for learning. In the specific case of short-snippet storytelling, all we can say as educators is that it is hastily gaining popularity among the under-30 age group and worth looking into.

Additionally, compelling the use of this method may be beneficial to learning, but may also not achieve intended goals. Because of its open-endedness, student stories may never touch on the intended concepts or experiences. Indeed, the broader toolbox of media in short snippet stories can be helpful though as it provides more ways to share stories. As such, it may also generate better levels of interest and motivation as short snippet storytelling doesn't privilege one form of media over others. The form is in the stringing of stories, not in the media used to convey information. Everyone can be a storyteller. These combinations of media encourage experimentation, critical thought, and expression in the telling of stories.

Collaboration in Cloud-Based Writing Apps

4

Google Docs is for more than just personal writing!

Writing can be really boring, but is one of the most necessary things in which every person should gain mastery. Over the last 5 years, **cloud-based writing services** such as **Google Docs**[1] have turned boring, old word processing on its head, making it much easier to be productive and work collaboratively. How, though, can teachers leverage cloud-based writing apps such as Google Docs to make the most out of writing? You'd be surprised...there's more to Google Docs than just 12-point font, double spaced text.

The beauty of one document

For many new to Google Docs and other cloud-based writing services such as **Microsoft Office 365**[2], it's a surprise to learn that

---

[1] App- Google Docs: https://docs.google.com/
[2] App- Microsoft Office 365: https://login.microsoftonline.com/

there's no actual file saved on a disk. Instead, it lives on a server and is accessed via the Internet. In the case of cloud-based writing, there is only one version of your document. The value of only having one document is that this one version can be accessed from many different devices and by many different people simultaneously. In this article, I discuss Google specifically. However, there are many other cloud-based writing and collaboration apps out there, especially for education. A quick google search[3] reveals many apps out there to help students write, collaborate, and develop positive writing skills.

Google has made some significant improvements to Google Docs in recent years. Primarily, you can now work on documents offline, which used to be the primary challenge with cloud-based services. You can open or sync working files to your device when you have an internet connection, and when your computer or mobile device is not connected to the internet, you can still work on the document. Additional improvements include better security and administrative capabilities for sharing documents. The sharing controls are robust and you know when and who accesses certain documents.

In addition, Google Docs makes it easy to reverse documents to previous versions if the group doesn't like the newest revisions, or someone decides to be a clown and mess up the group's work. In the cloud, work is rarely lost. Each change becomes an entry on a list from which you can go back and forth on edits.

With these features, cloud-based services give new life to boring old word processing tasks. Tracking changes and versions of documents, or emailing word docs back and forth becomes obsolete with one, single document that an entire team can access. Plus, backups become a redundancy, as the Google Docs app continuously backs up your work.

---

[3] For example, I tried the search terms "education writing apps" and it turned up a number of writing apps and services geared toward student use.

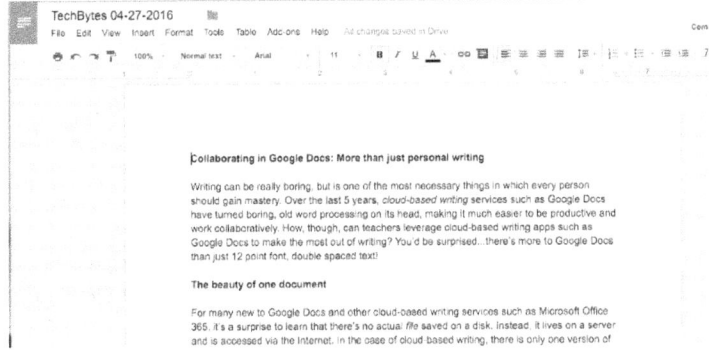

*I wrote this chapter in Google Docs—
It has become an important part of my daily workflow!*

Thinking pedagogically about Google Docs in the classroom

As reflected in my other chapters, I like to challenge myself to think about not what a technology does, but instead what activities it will help you do. Thinking in this way helps develop a perspective about the usefulness of specific techs in particular contexts, such as classrooms and personal work. To that end, here are some activities that Google Docs helps facilitate in the classroom:

Same-time group writing. Perhaps the best feature you get with Google Docs is that multiple people can work on the same document at the same time. This really opens the door for collaborative work. Teachers who think of some creative activities for students to do that align with pedagogical principles can really take advantage of synchronous writing. Students can watch in real time as things are typed and provide feedback. Students can each work on separate parts of the document and then share what they did. Groups can negotiate edits to text in real time. Concepts can be explored and defined via the web and "put onto paper" as groups come up with shared definitions and understanding.

Asynchronous group writing. Similar to the synchronous writing capabilities discussed above, asynchronous writing extends opportunities for working together outside of the classroom and

The beauty of one document

For many new to Google Docs and other cloud-based writing services such as Microsoft Office 365, it's a surprise to learn that there's no actual *file* saved on a disk. Instead, it lives on a server and is accessed via the Internet. In the case of cloud-based writing, there is only one version of your document. The value of only having one document is that this one version can be accessed from many different devices and by many different people simultaneously.

Google has made some significant improvements to Google Docs in recent years. Primarily, you can now work on documents offline, which used to be the primary challenge with cloud-based services. You can open or sync working files to your device when you have an internet connection, and when your computer or mobile device is not connected to the internet, you can still work on the document. Additional improvements include better security and administrative

> Jeremy Riel
> 9:35 AM Today   Resolve
> You can have a full conversation about the document in the comments window!
>
> Jeremy Riel
> 9:36 AM Today
> And reply to comments as they occur.
>
> Reply

I find myself having full conversations in the comments bar when working with others. The sidebar becomes a document within the document!

expanding classroom time. Even if students are not online together at the same time, they can benefit from others' participation on compiling a document. Google provides features to show "what's changed" since the last time a person logged in, which allows for easy catch-up. The same actions for discussing text, arguments and concepts in the context of a written text are facilitated by Google Docs.

Merging individual contributions. In Google Docs, you can either edit text directly or "suggest" changes. Select the "suggest changes" option in the top right corner of the screen when you have multiple collaborators on a document. You can see the revision history of the document in under the "file" menu and by selecting "view revision history." Individuals' contributions to the document are color-coded in the document's history. These contributions can be compared by collaborators, allowing the group to decide what elements to keep and what to discard. This is also a great opportunity to develop shared understanding on concepts, arguments, and writing structure.

Comments and chat window—a "document within the document." The comments function in Google Docs is a strong tool for collaboration. I like to think of the comments sidebar as the document within the document, offering a channel for people to discuss the content being written. Comments can be offered on specific sections of text, which are highlighted. Comments can also be made on comments, which allows for side conversations to happen without ruining the original text. When comments are no longer useful, you can mark a comment as resolved, and it disappears from the comments sidebar. All comments, however, are saved in the "comments" button at the top right corner of the screen.

Similarly, there is a synchronous live-chat window that can be activated when more than one person is editing a document simultaneously. This chat window can help collaborators organize their work and discuss issues that arise. These tools are useful to take advantage of!

Notifications on comments and edits. Google Docs provides timely notifications for documents. As activity can be asynchronous, it can be difficult to remember to check back on a document often enough. With G-Docs, you can get notifications sent to your email whenever people do things in a document. This can be anything from minor edits to offering comments on the document. You can also email all collaborators directly with the "email all collaborators" function, which helps improve communication and organization when working on a document together.

Add-ons, plugins, and other apps to extend functionality and activity. Google Docs comes with a library of extensions that you can add to the app. These are usually developed by third parties to give users additional functions. Access the library of add-ons by clicking "Add-ons" in the menu bar and selecting "Get add-ons." Some add-ons are as simple as the ability to insert special characters, to as complex as organizing to-do lists and other collaborative tools that don't come "stock" with Google Docs. Personally, I use add-ons in Google docs to help me insert math equations, special characters, and to format tables in more custom ways. You can also build charts directly in Google Docs using the "chart builder" add on. In addition to add-ons, you can integrate Google Docs with **IFTTT**[4] (short for If This, Then That), a service that allows you to link apps you use to your docs to automate tasks and provide reminders. The combinations of app integrations on IFTTT are seemingly limitless with Google Docs/Drive[5] and can really improve the experience of users based on their interests.

---

[4] App- IFTTT: http://ifttt.com/
[5] For examples of integrations, check out https://ifttt.com/google_drive to see what other IFTTT users have done with Google Docs & Google Drive.

5

Digital Demonstrations Have Become Quite Sophisticated!

The educational value of capturing what's on your screen

We've all endured it before. Tech support via phone or written instructions can be one of the most excruciating experiences in humanity. It's eternally frustrating to follow esoteric, repeated instructions to click buttons that don't seem to exist, having to find elusive menus, and be directed by enlightened tech gurus while we feel helpless to intervene on our own behalf. How-to guides can only go so far when demonstrating visual elements with words (with this article equally guilty!). Describing and talking about apps, software, and computing devices can quickly cause eyes to roll and frustrations to rise, especially when working on class tasks. Fortunately, these are also experiences of an internet era long past. What if we could actually see what's being described to us on the screen?

Wait….we can? Well, of course we can, or else I'd have nothing to write about!

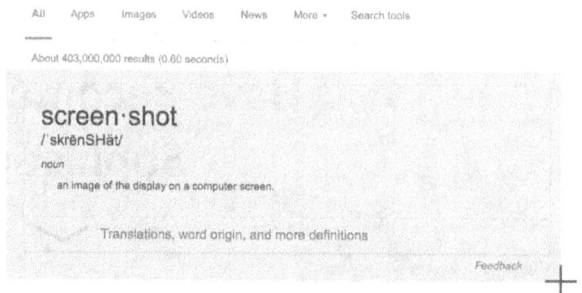

A screenshot of a Google's definition of screenshot!

In this chapter, I'll share a bit about my thoughts on the types of screencapture technologies in 2016 and how we could use them toward principled digital pedagogies. These screencapture technologies allow both the sender and receiver to illustrate what they are both seeing and doing, increasing the richness of communication for both sender and receiver.

Our culture is now dependent on the screen. For most tasks, in order for us to interact with the device, we have to use the flat screen as a window to the digital world. Despite many calls to other ways for us to interact with our devices (such as speech or gesture) our technology-based lives are dominated by the screen as the interface between us and the devices we use. To best convey ideas and concepts, it's up to us as educators to communicate with these screens in more ways than with just words. Screencapture allows us to add an additional layer of information to our communications by adding the visuals of the screen interface itself. Our thinking can be visualized and tracked as we navigate on the screens of our apps.

Thinking Pedagogically about Educational Technology 2016 41

A screengrab of me writing the paragraph below in Google Drive. Screengrabs can demonstrate how things work, illustrate one's thought processes, or show how to do tasks. (PS, I used Screencastify for Chrome as I wrote this article on my Chromebook. I used Giphy to make an animated gif image of my writing, which I posted on my blog. But, for this book, a flat image will have to do ☺).

When working with digital devices, I'd say a picture is worth all the words you can muster. Screencapture technology provides the ability to record still photos and videos of the actual interactions on a screen as they play out, which provides additional ways for communicating ideas about things we see on the screen. Visual demonstrations let you see the process and not just talk about it.

Technologies for screencapture:
Show me what you see

When "screengrab," "screenshot," or "screen capture" is said, there are usually a few different methods being mentioned, depending on the type of screen being grabbed (e.g., mobile, desktop) and the timing (e.g., still shots, prerecorded video, live video). Each of these methods are discussed in the sections below, along with links to some good apps for doing screengrabs of your own.

Screenshots: The still images of digital life on the screen. Taking screenshots is the first step toward demonstrating what you're seeing, intending, and thinking related to the screen interface. In 2016, "screengrabs" are no longer considered wizardry and numerous tools can be used to take, edit, and annotate screenshots. On a PC, one simply has to hit the PRTSCRN or "Print Screen" button on the keyboard, which puts the image on the clipboard for copying and pasting. On a mac, one has to push three buttons together to grab a screenshot: CMD+SHIFT+3. For mobile devices, each device has its own button combo to take a screenshot, but it usually involves holding down a combo of the power and "home" button on the phone for a short time. Most devices with screens today allow screenshots to be taken, which can be used as a powerful tool for communicating ideas within digital contexts.

Annotations of screengrabs are also really helpful. "Drawing" on screenshots or highlighting portions of screens are a great way to draw attention to screenshots and provide additional information. Photoshop and other production-level editing apps are no longer necessary, as educators and students alike can quickly edit and annotate these images within in-browser editing tools like **Pixlr**[1]. give it a try! Pixlr is simple to use and offers all the basic editing and text annotation functions one would need for quick screenshots.

Action screen capture: Show me what you're doing, don't just describe it. My favorite screengrab software on the free tier is called **Screencast-O-Matic**[2], which I often point my friends to when they need a quick grab. This screengrab software that anyone can use in-browser. For most laptop/desktop users (e.g., PC and Mac), it can be turned on as soon as arriving at the website—just push record and go! There's time and space limitations on what you can record,

---

[1] App- Pixlr: https://pixlr.com
[2] App- Screencast'O'Matic: https://screencast-o-matic.com/home

so it is ideal for quick jobs to quickly illustrate concepts or provide a demonstration of on-screen procedures. Unfortunately, Screencast-O-Matic doesn't work on a Chromebook, but there are similar Chrome extensions that you can download and use (for instance, **Screencastify**[3] and **Capturecast**[4] work pretty well).

For a higher quality screenshare, **Snagit**[5] is a wonderful choice. This software has a retail price of $50, although there are great educator discounts. This is a well-designed piece of software that will up the quality of your screengrabs by allowing you to record many types of media, including your voice over the video. Snagit also allows users to annotate videos by placing text, images, and drawings over the video feed. This allows users to provide additional information and context to their screengrabs.

Although it has a steep price point, **Camtasia**[7] is the full-production suite of choice that can be used to capture screen activity—but may be beyond what most people need. For high quality screensharing productions, such as instructional videos, Camtasia allows users to 1) grab high-quality videos of whatever happens on the screen (and in any screen size) and 2) output these videos in formats that are useful for today's purposes via what's called a non-linear video editor. For those readers who haven't dabbled in video production, non-linear editing allows you to add many streams of information to a single frame, including images, audio tracks, video tracks, transition effects, and text. If you find yourself using screen captures often for demonstration and want to up your production quality, Camtasia is the gold standard. It's the one I use all the time for my work.

One of the challenges in finding a good screencapture recording software is the ability to have it be in a usable format for today's

---

[3] App- Screencastify: https://chrome.google.com/webstore/detail/screencastify-screen-vide/mmeijimgabbpbgpdklnllpncmdofkcpn?hl=en

[4] App- Capturecast: https://chrome.google.com/webstore/detail/capturecast-chrome-screen/dmhhfoemgdlphenmfoicajbakonjcgee?hl=en-US

[5] App- Snagit: https://www.techsmith.com/snagit.html

[6] App- Snagit educator discounts
http://shop.techsmith.com/store/techsm/en_US/pd/productID.289742700

[7] App- Camtasia: https://www.techsmith.com/camtasia.html

video streaming standards. Unfortunately, Flash and AVI are two common output formats for many screencapture software, which are increasingly incompatible with web browsers and video streaming services. A google search will reveal many competing apps for grabbing screen videos. It's important to remember, though, that it's almost as important that your viewers can see the end result! In other words, make sure you pick software that outputs videos in a commonly usable format in 2016 (e.g., .mov or .mp4)! Most video capture software also have free trials, so educators should make sure to take advantage of a trial to make sure it works for their personal and students' needs.

Live streaming—let's be on at the same time. Screen capture also goes beyond show and tell. Collaborative work can benefit from live streaming just as much as teaching and demonstration. Take, for example, when a group of students works together on a project together on Google Docs: everyone can see the same screen and can communicate on how activity can proceed. Because attention is focused on the screen, the very presence of the images in communication can improve the quality of work. Students can then focus their communications on what they are seeing, making work much more effective.

Screensharing is popular in other live contexts today, such as webinars and conference calling. Focusing visually on the screen can help drive conversation and improve the quality of work. Another interesting example today has been with increase in popularity of live video gaming and e-sports. Gamers share live videos of their games on Twitch, YouTube, or other sites to vast audiences who enjoy watching (and even playing along) with these new stars of live TV. There's likely something to be said about the live focus on a visual element for learning and engagement. The cool thing is that our technology has finally caught up to the point it doesn't require $20,000 in specialized equipment and dedicated network connections to do so.

Mobile: the screen is now in our hands. Over the last decade, the screen has left our walls and found its way into our hands. With more people using handheld mobile devices as a part of their daily routine, it has become important to demonstrate skills, procedures, and concepts in the same terms as the device's graphical interface. In other words, it's much easier to show and do with a mobile device than to talk about or describe using "just words." The device is built predominantly around visual elements. Thus, screengrabbing mobile devices has become a useful tool for teaching others how to do mobile-based tasks, for learning how to use mobile apps, or for integrating mobile tools into everyday work.

However, the challenge with sharing the screen of a handheld device lies in the diversity of these devices. Because everyone's device is different, there's not one app that will work for everyone's device. Having tried out screencapture apps on many devices myself, I have some that I recommend. The simplest way to live-broadcast your screen is to "mirror" or "cast" a mobile device to a larger screen, such as a projector, TV screen, or computer. This is done easily today with apps such as **Chromecast**[8] and **Apple AirPlay**[9] (AirPlay comes installed on all Apple devices). With these, you can connect to a compatible device that is plugged into a monitor, screen, or projector, such as a Chromecast attachment or an **AppleTV**[10].

In terms of apps that "mirror" your phone's screen on a computer (i.e., you see the phone's screen on your computer monitor), **Reflector**[11] comes with my highest recommendation. I couple this with Camtasia or Snagit to put the live feed of the phone's screen on my laptop computer, then screengrab that feed from the computer screen. This is how I do demonstrations on how to use mobile apps for the faculty at the UIC College of Education in my day job. In addition to Reflector, **AirParrot**[12] allows you to share media across screens that are connected on the same network, which could be really useful in a classroom setting (think of sharing windows between students' phones or computers as if they are all

---

[8] Hardware- Chromecast: https://www.google.com/intl/en_us/chromecast/
[9] App- Apple AirPlay: https://support.apple.com/en-us/HT204289
[10] Hardware- AppleTV: http://www.apple.com/tv/
[11] App- Reflector: http://www.airsquirrels.com/reflector/
[12] App- AirParrot: http://www.airsquirrels.com/airparrot/

connected to the same computer). To record the screen activity directly from a device, **ScreenStream Mirroring**[13] is an app for Android that does what you want. Unfortunately, Apple does not allow direct screen recording on their devices, so the Reflector/Snagit combo that I mentioned above is the best way that I've found to capture Apple device interactions.

Thinking pedagogically about screencapture

If we take a step back to examine today's screensharing technologies, we can find a few valuable pedagogical affordances that aren't really available with other tools. Some of these affordances can support some fun projects for students, so it's worth trying out some modern show-and-tell tech in the classroom.

It's technology for demonstration....with some muscle. You get a bit more bang for your buck with screengrabbing tech. Instead of just using words to describe skills, procedures, and concepts, you can *demonstrate* what to do, step-by-step, *live.* Visuals of what is happening on screen give discussions some extra oomph. This is particularly valuable in cases where specific software is being used or skills are needed (e.g., internet searching/research, writing strategies, and even communication or collaboration with others). Really, any screen-based activity could benefit from screensharing technologies!

Collaborate and work alongside one another. Collaboration is a key component of many educational activities. Screensharing promotes collaboration by default whenever work is centered around a computer screen. Often, screens are individually focused activities. By sharing screens in live contexts (such as with Google Docs), students can work together and see what others are doing. This adds a significant amount of richness to activities and makes it

---

[13] App- ScreenStream Mirroring:
https://play.google.com/store/apps/details?id=com.mobzapp.screenstream.trial

so participants don't have to describe what they're doing when they work with others.

Annotation. Additional layers of information on top of the screenshot or screengrab video give a lot more information than the screengrab itself gives. Modern screengrabing technologies allow participants to include voice notes, text notes, graphics/images, and drawings to their screengrabs to make the information they contain much richer than the screen itself. I find myself noting on screengrabs all the time to indicate to others where to pay attention, things that are important, and jotting notes about things I see. If classes use screen-based technologies, annotation can be especially useful for communicating additional ideas in the context of the screen.

Authenticity. With screengrabs, we don't just talk about hypothetical situations: we show actual screens in actual use. It's much more effective at communicating screen-based ideas. It's also highly contextualized, making it valuable for use on projects that have highly specified steps or when trying to clearly illustrate procedures and skills for students. The authenticity grabs attention better because it is actually happening, and students can try stuff on their own with screengrabs as their guide.

Student screengrabbing—it's not just for teachers! Try getting your students to make screengrabs on their own! It's not just for teachers trying to demonstrate ideas and illustrate skills, but also for students to communicate with other students and with teachers. Students can use screengrabs to get help and more clearly communicate what they're doing on the screen. Screengrabs can also be useful for demonstrating, storytelling, and discussing on the students' end, making it a useful communications tool to leverage student creativity. The application of screengrabbing tools, including annotation, make use of many forms of rich digital media, giving rise to students' development of digital literacies and the ability to communicate visually. These are great skills to have in the 21st Century!

6
Everyone's Got a Bit of Librarian in Them

The pedagogical value of collecting digital media with archive and curation tools

How do we begin to keep track of the numerous resources available on the web when working on a research project? What happens when you want to send that perfect article that you found on Facebook the other day to a friend, but now you can't find it? Search dilemmas like these can be solved in part with the help of **archive or curation tools**—software that helps you sort and find web-based resources. In this chapter, I take a look at archiving and curation as activities for learning and the applicability that these activities have in the classroom.

This is one of those topics that are hard to write about generally. Customization is the signature feature of digital archiving techs and my method of archiving may not work for everyone else. Every morning, I read a bunch of headlines from websites that I follow and save the ones I like to read either right then, or to read later. While skimming my daily news and tech gossip headlines, I take part in a curation process that I've honed over the last 10 years

Cataloging information resources is a skill that is useful for everyone in today's digital world Photo credit: Megan Amaral (via Flickr-CC[1])

that works for *me*. There's no right way to make a collection of web links, videos, and notes. So, yes, everyone's personal digital library is different…and *should be* different. It's *your* library for *your* needs. As such, there's great value in introducing these technologies to students and having them create custom collections of resources that help them achieve their goals or express their interests in a way that is authentic: you have a constantly evolving collection when you're done that you can share with others.

Every now and then, some new apps come along to make this process easier. For me, I find a way that helps me organize the articles and other media that I find that I may want to use later for either a project, or just to have in my personal library for something that may come up. Web curation and archive tools are one of those that have meaningfully changed how I read headlines and save stuff for later. With petabytes of content generated every day on the web, we need a way to set aside the nuggets of useful information for us to retrieve later.

---

[1] Amaral, M. (2009). Card catalog 2. Flickr Photo. https://www.flickr.com/photos/mamsy/4175783446/

K-12 and entry-level college writing courses on how to do research and argumentative essays have long suggested strategies on how to keep research projects organized, such as how to record, sort, and properly cite resources like papers, books, and websites. However, the digital world has complicated matters as the amount of information has exponentially increased with each passing year. Indeed, there are more news articles than ever from every perspective, hundreds of op-eds and essays on any topic imaginable, and an ever-growing number of blogs and personal websites.

I'm not the first to write on digital archiving and curation as activities for learning, and certainly won't be the last. My writings, like the writings of everyone else, are doomed to get lost in the sea of internet info unless we have tools for grabbing things we find useful. There's so much information being shared that problems of finding information notwithstanding, we need a way to capture and retrieve information sources after we find them.

If you don't have a research project that you are working on, you'll still likely find some value in archival tools for curating your own personal collection of web links that you find fun or aligned with your hobbies or career interests. Everyone has a bit of a librarian in them, and these archival tools can be good for anyone with hobbies, interests, or things they like to organize for either themselves or to share with the world.

Tools for archiving digital resources

There are probably hundreds of archiving tools out there that require no programming skills to set up. I'll cover a few of the popular ones below, but talk about what distinguishes them in general.

The bookmarks, stars, or favorites in your web browser is one of the original archiving technologies that were available on the web. They continue to be available today, but they are difficult to tag and search and lack the robustness for research and sharing that modern tools have. However, bookmarks are still really useful for small collections of personal links that you visit frequently, or websites for which you don't need a lot of organizing. If you don't need to organize or search your bookmarks, then don't rule them out. However, they can get pretty unruly if you let your bookmark collection grow (as mine has done for the last decade...)!

Second, there are personal link-saving and link-sharing websites that provide the search and sorting functionality that is missing in conventional bookmarking tools in a browser. Websites like **Diigo**[2] and **Delicious**[3] are great tools for doing the one simple task we need: saving web links. But we get a bonus as well. Users can sort, tag, and put the links into folders, as well as choose which parts of their collection are public and private. The privacy controls coupled with search capability have made these a favorite of mine over regular web bookmarks for years.

Another bookmarking service that has gained a lot of popularity is called **Pocket**[4]. When you save something to pocket, it creates an easy-to-read version of the page in your Pocket for later ease of reading—and I have to admit, it's really nice. Searching and tagging are very good in Pocket as well, making it a great private repository of links. In addition, it links well with multiple devices, as Pocket can simultaneously live on your phone, tablet, and laptop web browsers. With a simple "save to Pocket" button, you can capture anything with a URL web address and view it on any of your other devices. As a result of its simplicity and that it can connect to other services like **IFTTT**[5] (short for "If This, Then That," a tool for automating tasks) and **Buffer**[6] (an app for scheduling links to be shared on social media), Pocket has become a central part of my daily reading.

Notetaking apps have also gained in popularity for saving links in an organized way. **Evernote**[7] is probably the most popular of the legion of apps available to take notes. Notetaking apps help you

---

[2] App- Diigo: http://www.diigo.com/
[3] App- Delicious: http://www.delicious.com/
[4] App- Pocket: http://www.getpocket.com/
[5] App- IFTTT: http://www.ifttt.com/
[6] App- Buffer: http://www.bufferapp.com/
[7] App- Evernote: http://www.evernote.com/

keep more than just weblinks organized, as you can take text, voice, video, and other notes, as well as take "clippings" of other resources, like a news article. It's a great option for people who don't want to limit themselves to certain types of media in their collection, but it's not easy to publicly share a folder (think of the bookshelf metaphor above) and publish it on the web. Evernote also has space and bandwidth limitations, which require you to move to a paid plan in order to be a high-volume user of the service. That being said, I really like Evernote and use it for notetaking. However, I use other services for saving and searching for web resources as If have found that they work better for my workflow.

Finally, the concept of the "pinboard" website has gained in popularity in recent years, which is a visually appealing way of seeing resources. **Pinterest**[8] is arguably the most popular app for saving web resources in this way. Pinterest calls itself "a catalog of ideas," with curators saving and sharing items of interest on what are called *boards*. Boards illustrate a variety of pinned resources that are related to a central idea. Pins are simply any kind of web resource that they deem worthy of being attached to the board. These boards are visually stimulating, as readers typically see multimedia content when looking at a board and its theme. Although Pinterest has gained a reputation for being only for wedding planners and foodies, it's an excellent example of a tool for curation and public sharing of resources. Although I've not personally used Pinterest, there are other examples that are similar to Pinterest that I've enjoyed. These include **Goodreads**[9] or **Libib**[10] for book or media library collections, or **Snupps**[11] for creating a board to show off a general collection of stuff (like my Star Wars toys or my Lego collection). Web resource archives don't have to resemble stodgy bookshelves or academic-like bibliography lists of resources—they can be pleasing to the eye as well!

---

[8] App- Pinterest: https://www.pinterest.com/
[9] App- Goodreads: http://www.goodreads.com/
[10] App- Libib: https://www.libib.com/
[11] App- Snupps: https://www.snupps.com/

Regardless of the tool you use, remember that it's primarily your library: make it work for your needs.

Curating your collection

The goal of archiving solutions is to consider what you'll need later. This isn't always intuitive at first, but the ability to organize your digital resources reflects the true power of these tools to make your and students' lives easier. In addition, depending on if you want to publicly share your personal library, organizing your resources will certainly help visitors find value in your collection.

Sure, it takes a bit more time to keep your stuff organized, but it's worth it. It's just like folding the laundry: I always prefer to pull out a nice bundled pair of socks from the drawer when I want. Those socks don't fold themselves, though (…grumpy face). The same applies to web-based resource organizing: services like Delicious, Pinterest, Pocket, and Evernote help us to capture and sort web links, videos, and files quite efficiently.

Someday, though, the laundry will fold itself—at least metaphorically. These technologies will get good enough pretty soon where they will be able to sort articles and web links into useful categories for us all by themselves. But, we're not there yet, so let's talk about how to sort.

So, let's get tagging! Many of the systems I recommend in the next section are based on the common "folder/tag" architecture. Simply put, we store resources (like an article we want to save) into *folders* (similar to what you do on a desktop computer) that we make so that we can browse all of the related resources that have likewise been put in a folder or attach tags to the article to help with searching for it later.

If a system uses folders, we can name them anything we want—that's where the custom curation element comes in. Think of it as your very own digital bookshelves, except you can arrange the books in any way you want and have as many shelves as you need. In many ways, it's just like the desktop computer folders in which you put documents. For instance, if I want to collect useful news on tech, I might make a folder called "tech news." If continue this line of thought of thinking of folders as bookshelves, we can browse the bookshelf for interesting things or relationships between the resources on the shelf. It's similar to what you can do at a physical bookstore or a library, examining the stacks and browsing the shelves for interesting things. This is what I've been calling the *browsing effect*, or the ability to find connections and generate new ideas just because resources were lumped together on the same "shelf." Archiving tools give us the power to collect resources and organize them in a way that gives us useful new ways of seeing our information. And the best part is that you can browse others' publicly available digital "shelves" as well.

...It's also way easier to find your things when they're put away on the shelf! :

In a similar process to assigning folders, we apply "tags" to links in some of these archive systems. We can affix as many tags as we want to describe the article, and typically the more we put the easier it is to find the article. Think of it as putting a sticker or nametag on the article, and that you can scan that nametag really easily when you want to find it in a pile. Back to the laundry example, imagine that each sock, shirt, and towel has a nametag on it. With a special, hypothetical mobile phone app that doesn't yet exist, what if we can cause nametags on clothes to light up if you search for them on the app, or better yet sift themselves up to the top of the pile? That's the principle behind tagging—we assign labels to resources so we can find them easier later by only searching for that label.

Don't be worried about how you are sorting/tagging articles when you get started. Some tags that are useful to you may only become evident as you've been doing it for a while. As such, your system can change over time, and it's perfectly normal! I often find myself creating new tags in my pocket/delicious archive.

To have a successful archive, it's important to just keep in mind how you want to structure your resources. I've found it best for me to just mimic how I sort things in my head—things that are news I label with a tag called news; things that are helpful to a research project I label with a tag that is specially named for that research project (e.g., "*#weeklyArticles*"—and I sometimes truncate or combine words to make a unique word that I can easily search for later).

Including a hashtag on a tag can also be helpful so you can find only your tags with that label, and not just any occurrence of the word in your database. For instance, typing *#news* would find anything that has that exact string of characters—which would only be your tags that you labeled *#news*. On the other hand, simply typing "news" without the hashtag would turn up every article that had the word "news" in it, regardless if you want to see it or not. Sometimes that's helpful, but the hashtagging system became popular for this very reason in online archiving and communications tools.

Another thing to think about when organizing your resources in an archive is to consider 1) whether having an audience would be fun or useful and 2) what that audience might want to see if you open your archive bookmarks up. As a technology used for education, the development of a public personal archive would be a nice project around a topic of study. Wikipedia is a great example of a public archive, with each article pointing visitors to many useful sources on any given topic. As Wikipedia is foremost public-facing, care needs to be taken to consider how the public will view the articles, evaluate whether they will find what they are looking for, and to provide information that they would find valuable in the first place. I have a public Delicious archive that I store articles that I think people who are into the intersection of education and technology would find interesting. If you have a public archive, it's worth considering every now and then what you're putting online and how your audience might interpret it.

Thinking pedagogically about archival tools

So, what are some of the pedagogical implications for archival tools if we dig a bit deeper into classroom application?

Knowledge creation. Readers who are fans of learning philosophy will see an immediate benefit of archive tools for assisting with knowledge creation and information synthesis among learners, especially when archives are social. There are many opportunities to see connections between items in a folder, board, or among a particular tag. When working with tags specifically, we can see how some tags are related to or contrast with others, causing us to do a bit of critical thinking and idea generation. I say "we" here because it's not only students who can learn from these tools when used in a classroom. Potentially anyone who interacts with the archive, including parents, teachers, and others from a student's community, can learn and continue to create new knowledge by sharing critiques and ideas about the archive.

Content-specific study and deep exploration. Curation is great for content-area study, giving not only a storage and retrieval space for resources but also a place for thinking and reflection. Are you studying modern American literature? You can create an archive for that to capture titles in your range of study, but also critiques, media, and articles on the places and themes of the literature. Putting these all together in one space gives learners a new perspective on their area of study via that browsing effect I discussed above. Studying weather in the real world? You can create an archive for news articles of weather occurrences, as well as articles and media on the science behind weather phenomena. Studying big social issues like crime or the economy? You can create archives into which articles can be sorted and retrieved for later use. I've done subject-based archives many times and find them to be helpful not just for organization, but generating insights as well.

The browsing effect and critical thinking. I discussed what I call the browsing effect above (I'm sure there are other names for it in the library and information sciences). What I mean here is what happens when you browse the stacks at a library, which, unfortunately, doesn't happen nearly as often today as we switch to dominantly digital media. Sometimes ideas pop up when you're walking the stacks. As you're looking at book spines along a row of similarly categorized books, questions seem to pop up about how things are related and why stuff is there. When I was an undergraduate, I would walk the stacks when working on political science term papers. I could see the various titles of the books and hypothesize on how concepts might be related. How was political economy related to decisions made by state government in a given decade? Two books of those topics close to each other often helped me make a connection via this browsing effect. Having an article archive helps me in much of the same way, especially as I write my weekly column. I often go back to my digital archives in Delicious and Pocket to get ideas on topics. The browsing effect is not a new phenomenon by any means—libraries figured it out years ago! But digital archives are bring back this valuable aspect of libraries that we are starting to lose by not visiting the physical stacks.

Current thinking capture. This one's simple: it's an archive of your thinking as well. Learners can look back and see what was found when, what articles influenced their thinking, what paths were taken to get there, and what sparked interests and questions at certain points. A reflection on one's thinking could be really revealing and provide many insights, especially when trying to garner personal lessons about one's growth. Students of all ages can benefit by asking questions of why they put articles in the bin and how these web resources are related and different. Digital archive tools capture our interests in a special way that even journals have a hard time grabbing. What outside influences helped us generate our thoughts at the time? We can get some of this from our archives.

Publication and creation. Finally, but maybe most importantly, digital archives give learners an opportunity to create work products for both themselves and for others. In effect, these digital archive tools give participants the opportunity to create a literal museum, library, or special collection that is unique—*they built it.* The archive itself has additional meaning imbued by its creator, making it more interesting than the individual parts of which it's composed.

Building a project like this not only promotes the critical thinking and subject-matter study mentioned above, but it also gives ownership to learners over an authentic activity. It's far more interesting and motivational to build your own set of resources and not be limited by the type of media in the collection. By allowing students to curate, it makes for a great project around a particular topic or learning goal.

7 Pokémon Go and Education

Finding lessons for learning in The Hit App of Summer 2016

In the Summer of 2016, the hit game that found its way onto the phones of everyone under 30 years old likewise found its way into the discussions of educators worldwide. Questions about whether **Pokémon Go**[1] can be used for learning should be expected as it gained popularity. Anything that is able to pull in such large numbers of players has have some lessons in design that can be beneficial for learning, right?

Although I don't think there are yet direct ways that educators could use Pokémon Go for broad learning objectives in a principled way, I do think that there's a lot we can learn from the Pokémon craze and find some inspiration for educational design. If we take a moment to think about the pedagogical possibilities from the types of activities in Pokémon Go, there's some interesting insights to be gained. What elements of Pokémon Go might be carried over to other designs and classroom activities? Let's check it out a bit further.

---

[1] App- Pokémon Go: http://www.pokemongo.com/

Pokémon Go has some interesting principles in its design that can be useful for educators. Above: A pokéball (a mainstay item in the game) gets the mortarboard treatment

What is Pokémon Go?

For the uninitiated, Pokémon Go is a game designed by **Niantic Inc**[2]. and **The Pokémon Company**[3] with which people play on their mobile devices while interacting in the real world. You can play Pokémon Go as you walk to school or work, or while you're sitting around with friends (or while at work). It's designed to be played for as little or as long as you want at a time and will work whenever you have a data cell signal and a working GPS.

In the game, you play the role of a *pokémon trainer* on a quest to catch small animal-like creatures called *pokémon*. Many articles have been written about understanding Pokémon since 1995, so I won't rehash old stuff here. The premise remains the same for the newest app that was released in July: players catch pokémon and train them to become stronger. The goal of the game is to become

---

[2] Company- Niantic Labs: https://www.nianticlabs.com/
[3] Company- The Pokémon Company: http://www.pokemon.com/us/

A Spearow on my street on the first day of the app's launch—as seen through the augmented reality lens. The Pokémon can just be "sitting there" in real life. It's up to you to catch them all and be the best there ever was.

"the best there ever was" and to catch all of the pokémon. However, with Pokémon Go, you catch these little monsters as you interact with the real world, which is the main departure from its previous iterations that were limited to a gaming console.

One thing that sets Pokémon Go apart is that it is an augmented reality app (see Chapter 9), in which you can "see" the pokémon that you're catching in the real world. It's as if you look through your phone as a special lens to see hidden things. It has been extremely popular with students of all ages: middle school, high school, and college. There are many pokéstops on college campuses and schools, which are real geographic locations at which you can restock on supplies for the game (specifically, the ever-important pokéball).

Millions of people between 15 and 45 years old have downloaded the app since its release in mid-July. You read that right, millions of adults have played Pokémon. The diverse age of the players has been a fascinating aspect of this phenomenon, as it seems that the Pokémon we have come to know over the last twenty years is not just a kid's franchise anymore. The designers of Pokémon Go have gambled on the nostalgic draw to the game. Although available to children, most people using the app are members of the generation that played the original games in 1995, which is now composed of adults that have discretionary spending. It seems the gamble has paid off, as some news outlets reporting that the app had generated over $250 million in revenue in the first few weeks of its launch[4].

I've played the game since the day that it came out to about the end of August, as everything about this game was fascinating to me. I've played in multiple states over the summer of 2016 (IL, CA, and OR) and everywhere I went I saw the same excitement for the game. Speaking of distance, the geographic differences and varying play strategies based on city type (i.e., between urban, suburban, and rural) are also making headlines. This will be especially important if we are to best understand the digital lives of K-12 and college students. Students in both K-12 and college have integrated apps like Pokémon Go as key components of their daily entertainment and social lives…at least for now.

As of writing this chapter in July 2016, here's no indication yet on how long people will play or whether this trend will die. Such is the case with any flashy, quickly downloaded app. All we know is that people *love it right now*. Some popular social apps explode in their use in the first months and then die immediate death. This was the case for the super-popular **Candy Crush**[5] and **Draw Something**[6] games, both experiencing severe drop-offs after some months of no new changes to the game. However, some games find the magic formula for sustained use over time. There are a lot of pokémon to catch and many new game elements that Niantic wants to implement in the coming months. Educators should keep an eye on Pokémon's growth and changes in interest from the younger generations.

---

[4] CNBC Staff (2016). Pokémon Go crosses 250 million in revenues since launch. *CNBC.* http://www.cnbc.com/2016/08/12/pokemon-go-crosses-250-million-in-revenues-since-launch.html

[5] Rushe, D. & Quinn, B. (2014). King Digital struggles following drop in popularity of Candy Crush game. *The Guardian.* 12 August, 2014.
https://www.theguardian.com/technology/2014/aug/13/candy-crush-king-digital-shares

[6] Tassi, P. (2012). Draw something loses 5M users a month after Zynga purchase. *Forbes.* 4 May, 2012.
http://www.forbes.com/sites/insertcoin/2012/05/04/draw-something-loses-5m-users-a-month-after-zynga-purchase/

Can Pokémon Go be used for learning?

Of course, the question on my (and every other educator's) mind is less of "what is Pokémon Go" and more "how can Pokémon Go be used for learning." The allure of using this popular app is just too much to resist! Anytime popular apps are used as hooks for education, care must be taken not to make it uncool by warping the activities too far beyond what made it popular in the first place. So, that being said, I think there are two ways that we can cautiously answer the question right now.

First is the content question: *what can be taught using Pokémon Go?* I think this question is still a bit of a stretch for us to determine solid answers right now. The education community has only seen the app in action for a month, so we are far from seeing principled use of the game for educational objectives. There are always some content areas to be explored and learned with any authentic activity in life, so invariably players will gain skills in some of these areas just by playing. These include:

(a) map and geographic literacies, as players are required to navigate their communities and new places using maps and geographic indicators
(b) data literacies, as players need to manage their pokémon training efforts and calculate the necessary resources for improving their status in the game
(c) math, as players perform a lot of operations on numbers in the game, often using their heads
(d) history, as communities are explored and landmarks are frequently visited during play

I am not sure if Niantic specifically included these as learning objectives, but they can be seen as bonuses to playing. For teachers that are interested in history, geography, physical education, and even some math skills, Pokémon Go can be beneficial.

But, I'm still convinced that Pokémon Go is a bit of a stretch to say it can be used to promote learning in deep, curricular ways. We just haven't seen any applications of the game yet in learning situations and it's all hearsay if it works or not.

Downtown Chicago is full of pokéstops – the blue and purple icons on the screen. In the game, you walk up to them and tap to get resources to play the game. Many of these stops are statues, monuments, plaques, and other community locations. You'll likely run into other players.

Pokémon Go's inspiration for ed-tech design (or, thinking pedagogically about Pokémon Go)

So, the second question related to applying Pokémon Go to education is "*what are the special properties of Pokémon Go and how can they inspire educational design?*" In my opinion, this question is more important right now, is more compelling, and frankly way more exciting. In the short time I've been playing, I've seen many elements of Pokémon Go's design that provide some insightful lessons for education.

It's important to preface this question with a statement on context in technology use. The act of playing Pokémon Go is the *context in which these activities take place*. They won't necessarily transfer over to another context, such as a curriculum or learning environment. The combinations of activities that players do in the game are done to achieve only the goals of the game—nothing more. However, that being said, there are some lessons to be drawn from the rapid uptake and intense passion of the player communities that has formed since the launch of the app in July.

Place can matter. In terms of lessons, Pokémon Go's greatest contribution is probably an argument that place can matter and can be integrated into today's digital apps, but only if there's a good reason to do so. The distance between players and gyms, the length players must walk to "hatch eggs," and the location of pokémon spawn points, pokéstops, and gyms are all meaningful, location-based aspects of the game that get people to move their physical bodies to other places in the real world. Games are played outside, in fresh air. What we're seeing is a weird fusion of digital gameplay superimposed on physical location (which is what we mean when we say augmented reality, right?). But people are only moving and seeing the sights because there's a reason to do so?—?it's required to advance in the game. Pokémon gives us insights into what it takes to get people to move around, get outside, and how play and learning can be reconnected to human movement instead of taking place solely behind a stationary screen.

Prioritizing place in activities may actually have unanticipated consequences in some areas, though. Stampedes have occurred in dense areas like New York and, as in the video below, Taipei, Taiwan, as trainers rush to find rumored rare pokémon as they briefly appear "in the wild."[7]

The videos of these stampedes are a bit comical. They do show, however, that physical location can matter to digital activities and alter other activities (such as driving or commuting). This is important because digital gaming is a domain in which activity was

---

[7] 王亭懿 (n.d.). [Pokemon go] Snorlax Gotcha! YouTube Video. https://www.youtube.com/watch?v=MoYjVTbLWyo

thought to exist only in cyberspace, and Pokémon shows us otherwise now.

Social aspects also matter. The original Pokémon game was mostly a solitary endeavor; kids grinding away at finding pokémon on a Game Boy was a common sight for parents to see. But, all the way back in 1995, it was interesting that Pokémon observers could see a number of social ways to play the single-player game. Catching them all was meaningless unless you could show your friends. And show off everyone did—and continues to do—with Pokémon Go. Just ask any player today and they would be happy to show you their collection of powerful pocket monsters. As such, the game is social. In the original game, you could trade your pokémon with friends or do battle with your most powerful creatures. Today, you can battle with friends at local gyms, hike together to find pokémon, or sit with some friends at a pokéstop with a lure in place to have a higher rate of finding pokémon. How Niantic has employed the social element of play is a big insight to draw from the app.

Nostalgia can be really powerful. As a social phenomenon, Pokémon is fascinating to me because of all the adults who play. People who grew up with Pokémon are returning to the franchise. I think this is because they are largely drawn by the nostalgia. Perhaps it's the Pokémon app that they always wanted: to actually be the player, physically moving through the world to catch and train Pokémon on their own. It's as if the technology finally caught up and hooked players with nostalgia. We should keep our eye on how we can use familiar pop culture and entertainment in education, as it seems on the surface to have had a powerful effect.

Maintaining variety and purpose among monotony can help with boring tasks (*or, alternatively, catch lots of rattatas for every dratini*). Another weird thing I've noticed is that the app has made boring, repetitive tasks doable and cool. Sure, you want to catch a rare pokémon. But for every rare one, you must catch 100 boring ones. How did Niantic get people to willingly do these repetitive, boring, and sometimes lengthy tasks? Perhaps it was for the promise of a reward, or an achievement. I do know that you need to catch many small pokémon to get "stardust" to increase the combat power of your pokémon that you use to fight at gyms. These victories are mini-achievements and a strong in-game purpose that keeps people playing. Your character also can increase its level when you catch hordes of boring pokémon, so there's probably something to these outcomes other than just for the sake of earning a badge or level. Players have a purpose with specific goals, and they do these monotonous activities to meet them.

Encouraging movement and getting outside is not that hard. Contemporary thinkers about kids in the digital age love to gripe about how people don't do anything and just sit around. The level of kids' physical activity is indeed a concern of parents, schools, health professionals, and policymakers alike. Pokémon Go demonstrates an important lesson about physical movement: we just need the right hook and a purpose to get outside. This is a good lesson of how all technologies are used for a purpose. Mobile games move us past the age of the console or desktop computer in which the only place to play was in front of a large screen. Maybe our lack of movement in our society is simply due to the extreme tethering we've had to the digital console, be it a computer screen, television, or mobile device. What if our activities and entertainment had purposes that were tied to place and movement? Pokémon is demonstrating some exciting things in this area. But don't drive and play! The app is starting to gently remind people that this isn't cool, with popup messages whenever your GPS clocks movement that is too fast.

Listening to the players is important. Much of what Niantic and The Pokémon Company are doing here is an experiment. Could they anticipate everything that was going to happen when they launched the app? No. However, they did decide to push the envelope and try a few new things to see what would happen, much to their surprise (and benefit to their coffers). Pokémon Go was largely built on an earlier experimental framework that supported a long-running social game called **Ingress**[8]. The iterative development of Ingress laid much of the foundation for Pokémon (and used a lot of its code!). The release and continued updates of Pokémon Go have created an interesting series of cause and effect between the small changes that Niantic has done and the subsequent response from the players. Although it's a good topic for another chapter someday, there have been many changes and subsequent bursts of anger from the players' community on these changes. The lesson for educational design is although the app developers have a basic set of goals and that they try to push participants toward those activi-

---

[8] App- Ingress: http://www.ingress.com/

ties, changes need to be sensitive to the community of users and respond to their needs and desires for it to maintain its high usage.

8
Making Sense of Minecraft For Learning

Can we succeed at bringing Minecraft to classrooms in ways that spark learning?

For a few years now, **Minecraft** has been the go-to app for kids. But, despite being around for a while, a lot of the game remains mystical for anyone over the age of 25. Much of the press around Minecraft has revolved around its educational possibilities. Indeed, Minecraft has many educational properties that, if properly applied, could really expand learning opportunities in many classrooms. I am excited about the positive attitude toward Minecraft. With new media and digital worlds that expand learning spaces to be online, we often fail to see learning benefits outside a game's intent to be a "math" game or "spelling" game. To see the value in video games for additional learning opportunities outside of their intended, programmed learning objectives, we need a new paradigm for connecting video games to learning[1].

In Minecraft, players actively change the world by collecting the very building blocks that compose the world and build creations with these resources. The game is played by virtually any kid aged

---

[1] Gee, J.P. (2014). What video games have to teach us about learning and literacy. New York: Macmillan.

A Minecraft classroom - Imagine meeting in-game for class!

2-12, and given a chance, kids would probably love to see Minecraft in a classroom. However, before taking that step, teachers should really consider what they want to accomplish with it. Minecraft has some wonderful affordances for learning, but to realize them, some critical decisions need to be made about what it will be used for.

I'd like to start this review by arguing that yes, Minecraft has many educational properties that, if properly applied, could really expand learning opportunities in many classrooms. It saddens me to see an immediate dismissal of using a new game, app, or activity that kids do today for educational purposes because it is "not for learning." To help along those lines, teachers need to know what activities students do in their everyday lives, including the types of games they play and with whom. Minecraft is exciting as it can be used as a hook to grab student interest or to leverage some of the in-game learning benefits it offers, such as being a simulated environment or a place for social interaction. Brave teachers who want to experiment with Minecraft need to know what they're getting into. Far too often, in-class game experiments become disorganized and fall off the trail in achieving the objectives they set out to achieve. To combat this, I believe the biggest challenge with preventing a flop is to make sure you know what you will do with the game. It sounds easy enough in theory, but teachers should really know what kinds of learning activities the game will support, and what pedagogies can be used to make the most of the experience. In short, we need to always think pedagogically about technologies we use.

In addition, it's also important to not "steal" all of Minecraft's fun aspects as it is applied to formal learning. Students might like Minecraft in an educational setting, but there are some things that they may also want to keep separated between their entertainment and educational lives. Teachers risk making it "uncool" by changing the parameters of the game past the point kids originally wanted to play it. To this point, I unfortunately don't have any answers - this will require some additional research over the next few years.

The discussion below is intended to help us to think pedagogically about Minecraft and what aspects of the game might really transform learning in formal classroom settings.

What is it?

Basically, Minecraft is a game that participants simply explore and build whatever comes to their mind. It's entirely open-ended. The game's primary characteristic is the *block*—think LEGO bricks as the basis for a computer game. Everything in the game is composed of block-shaped objects that compose other creations with these blocks. From expansive mountaintops to oceans in the game world everything starts as a block that represents a 1x1x1 meter area. Blocks are the atoms that make up the Minecraft world. Participants control an avatar character (called a "skin") as they interact with the vast (and seemingly endless) world to collect resources and build structures. Structures can be as simple as a small farmhouse to as grand as a ancient cathedral. Resources are everywhere in the game, with each object in the game represented as a "Minecraft block." Blocks come in a variety of types and contain an equally diverse set of properties for use in building, "crafting," and in-game play.

A grand cathedral, made of individual Minecraft blocks

Interestingly, it is a game with no "win state" or predetermined path. The only rules in the game are some simple rules of physics (e.g., you will fall because of gravity, and so will some Minecraft blocks – but only some follow these rules!). Players make the rules as they go along, limited only by their creativity. When playing with others, small-scale social rules tend to emerge, adding an additional layer of complexity to the game.

An exciting part of the game is that it can be played with others. It's not as much fun to build your creations by yourself and not have anyone share in what you've built. Multiplayer mode on *Minecraft servers* allow for people to play together. This brings a real collaborative element into the game as well, as people can work together to build or explore. It can create a competitive element too at times through competitions, or as kids can be sometimes, preventing attacks on one's builds. Because it is multiplayer, some care should be taken in classrooms to prevent "griefing," or the deliberate attacking and destruction of others' creations by bullies.

Imagination knows no bounds in Minecraft. Of course you can construct a football field in the ocean if you want to!

As such, the educational aspect seems like a no-brainer. Kids are really excited about Minecraft, spend hours playing it, and it seems to be a no-violence, creative game. Why wouldn't we want to leverage this tool for learning? In fact, Microsoft has recently been promoting a new special edition of the game for educators, Minecraft for Education[2]. The potential for learning has reached the ears of the Microsoft developers and others who are now building lesson plans centered around the game. Although there are specific benefits for bringing Minecraft in a classroom, we want to be careful about how it is used. Kids' interests in the game still need to align with what teachers are trying to guide students to learn. As such, some specific decisions need to be made by teachers looking to integrate Minecraft.

---

[2] App- Minecraft for Education: http://education.minecraft.net/

"Minecraft Steve" from my personal LEGO collection. Minecraft is now officially in LEGO form, as well!

Thinking pedagogically about Minecraft

There are many educational affordances that Minecraft brings to the table. If we think about the types of pedagogical activities that you can do in the game, it becomes clear that Minecraft gives students many opportunities that are not necessarily available in a conventional classroom. Although there are certainly more, I discuss five of these below.

Engagement. The adventuresome game is highly engaging. Minecraft is one of the best cited examples of Csikszentmihalyi's state of flow[3], a psychological state in which participants are so wholly immersed in what they are doing that time itself seems to become meaningless and intense cognitive effort is expended on the activity. Because players pour their entire attention span into the game, amazing things can be completed, and as a result, significant learning can occur if the activities are tied to some kind of objective.

Creative exploration. Creativity and open-ended problems define Minecraft play. Although there are some hard-coded physics rules in the game, most of the rules are defined by those playing.

---

[3] Csikszenmihalyi, M. (2004). Flow, the secret to happiness. TED Talk. https://www.ted.com/talks/mihaly_csikszentmihalyi_on_flow

This, by definition, is an affordance for creative behaviors. In addition, players are encouraged to explore, hypothesize, test, prototype, and interact with the environment. These aspects can be used by teachers to create places for creative inquiry activities mixed with creative expression. Because of the multiplayer element to the game, collaborative inquiry and learning could also be realized, as well as collaborative creative works. Given a few parameters to guide their work, students can create some amazing things in the game!

Customizability and personalization. Everything about the game is customizable. The character who is played (the avatar) can be customized with "skins," which are custom pixelated representations of the character. The whole premise of the game is to customize your environment: you build houses high up on mountaintops or in the sky, or dig deep into the ground to form underground lairs. Since there are no rules for play and no path to go, the only way to play is to craft and interact with what you build! Personalized environments give students ownership over their work and help them feel like they are a part, which can increase motivational aspects like interest and self-confidence.

Simulation. Students can create replicas of places or things that are important to them or their research and interact in these worlds. Many minecrafters have created historical reenactments of famous events or places, which allow participants to interact in these simulated worlds. Remember, the key here for Minecraft's popularity is not in its resolution to the actual thing it's representing, but instead the affordances for creativity that the game provides. It is popular because Minecraft lets anyone as young as two years old to create objects either from their imaginations or from historical or current events. Engineering and design principles are at the forefront of Minecraft designs. Planning your builds helps, but is not required because players can experiment with their designs as they build them. Because it is relatively easy to build and provides a space for participants to interact with their character avatars, it creates a new realm of relevance that can extend learning past the classroom.

As a classroom extension. Perhaps most exciting is that Minecraft is a virtual world that lives outside of the formal classroom. Learning activities that are embedded in a Minecraft world can be used to extend learning opportunities to outside of the classroom.

This could blend with students' entertainment activities, but may also create new times and contexts for learning that are impossible to achieve inside of a classroom alone. Playing with friends during their off time or building within the game outside of the class can increase the time available for learning activities by making it a fun, problem-based activity.

Try it out!

Try Minecraft out for yourself! For an online, multiplayer game, it is relatively cheap (approximately $25). It can be downloaded from the Minecraft website[4] and can be run on PC or Mac. There are also tablet and mobile device versions, although they are more limited than the PC or Mac versions of the game. Each version allows for you to join others' games, as well.

Keep in mind, however, that setting up a game for multiplayer play requires creating a Minecraft *server* on the Internet to host the game, which can take a few advanced skills. There are many services though that have popped up recently[5] to help streamline this process and make setting up a server super easy. There are also many walkthroughs on the Internet for setting up your own computer[6] to be a server for a Minecraft game. Many Minecraft "builds" can also be downloaded from various places. You can do a Google search for Minecraft builds and games, or for servers that host "builds" that you can join.

---

[4] App- Minecraft: https://minecraft.net/
[5] There are many services on the web with which you can set up a Minecraft Server. A quick Google search for the term "Minecraft server hosting" turned up a bunch of host services:
(a) MC Pro Hosting, https://mcprohosting.com
(b) Minecraft Server, https://minecraftserver.net/
(c) Apex Minecraft Hosting, https://apexminecrafthosting.com/
(d) Bisect Hosting, https://www.bisecthosting.com/selector.php.
[6] Service: Setting up a Minecraft server:
   http://minecraft.gamepedia.com/Tutorials/Setting_up_a_server

9

Augmented Reality in the Classroom

New realities for teaching and learning

Digital technology gives us many opportunities today to envision and create experiences that are not possible in real life. Through computer graphics, Hollywood blockbusters give us avatars, aliens, and avengers who can use superpowers to save the planet. Video games take us to faraway lands to complete quests for gold and glory, or build expansive Minecraft villages. But what if you could bring those experiences into the real world?

Could you bring the virtual world to the classroom? What if the physical space of the classroom could be used to bring digitally created elements to a student's experience? What if we could bring the endless opportunities for digital creativity into our everyday embodied experiences?

Augmented reality (AR) gives us tools to superimpose digital elements on the physical spaces in which we interact. Using a mobile phone or tablet as a "window" into the digital world, AR apps give people richer experiences by bringing digital objects to life. Humans can interact with them, change them, and learn from them. In short, by allowing objects to be in the same spaces as us, AR provides memorable experiences that are not normally doable via just a computer screen.

The term "augmented reality" often gets confused with "virtual reality." Although they are both similar, today's AR tools give you the capability of creating digital objects in the spaces where you interact every day. On the other hand, virtual reality (VR) refers to completely digital environments, such as video games, that are not a part of our physical environment. In VR, you typically wear goggles and other devices. With AR, you just use the phone in your pocket or tablet in your bag!

For instance, while walking down a street, you could hold your phone up to a AR-enabled building or street and get details about where you're at. You could hold your phone up to an AR-infused statue or monument and access valuable information about it. And in some instances, you can play games where the imaginary and unreal become real. A great example of this is Google Ingress, in which you work through a city with other players to capture other teams' "portals" of energy using actual physical locations for the context of play (which, of course, are displayed via AR on players' phones).

For those who play or know anyone who play **Minecraft**[1] (see Chapter 8), **Microsoft's Hololens**[2] has some promising AR capabilities for interacting with Minecraft worlds. Imagine placing your creations in three dimensions in your living room or classroom!

Using the devices that are already in most students' pockets, teachers can build AR experiences using free or low-cost tools. Most of the AR tools today are easy to use and can be deployed in most spaces with relative ease. Most of the heavy lifting on the technology side has been done by app providers. All that is required for classroom applications of AR are imaginative ways of thinking of the world and ideas for representing concepts that are being taught. AR even provides us with a new concept of what it means

---

[1] App- Minecraft: https://minecraft.net
[2] Hardware- Microsoft Hololens: https://www.microsoft.com/microsoft-hololens/en-us

to "learn something in class"—when that something is *in class*, what new pedagogies can we be inspired to try?

Thinking pedagogically about augmented reality

Instead of thinking what specifically we might teach with AR, it is sometimes helpful to first think about what types of activities AR lets us do. Below are a couple of the key advantages teachers can gain when thinking pedagogically about AR:

Can "give form to" stuff being studied. Despite having the ability to model phenomena on computers and use language to describe it, we still use our primary senses of sight, sound, and touch. Although AR can't yet make touch a reality, it can be used to create other sensory interactions as someone does activities in a room. AR makes things visible in our space by allowing digital objects to "be there" in the room with observers. As a result, AR can help represent complex things that are hard to understand from a static image in a textbook or website.

Imagine seeing molecules in your hands, or in the hands of one of your lab mates. Imagine seeing weather patterns, other places, or priceless artifacts in the classroom. Imagine being able to draw networks between people objects to see relationships, in real life. Imagine being able to give any idea a shape with which students can interact in the classroom. AR presents a new way of thinking about complex ideas and how we interact with them by leveraging space and the embodied experiences of people to create new depths of understanding.

For example, The IKEA furniture company has been doing this for the past few years by creating a virtual catalog in which you can virtually place furniture in a room. Pick a piece, hold your phone up, and *voila!*, you can see the potential piece of furniture where it would sit if you had bought it. No purchase necessary, no moving furniture around. New loveseats can be animated on the screen, using your room as a backdrop. This allows you to imagine how spaces may look with new furniture, but also can spark your imagination when objects interact in physical spaces.

Can give additional meaning to space. Sometimes, place matters too. Meaningful distances between objects, or even just spacing things apart can help students to compare and contrast ideas. Couple this with the ability to represent phenomena as objects in the classroom via the nifty mobile phone AR viewfinder, and you give students some powerful tools for analyzing and discussing ideas. As classroom space is also inherently social, you also gain tools for collaborative work. As someone who organizes a classroom, a teacher can also use space to orchestrate classroom activities. What if suddenly each corner of a room has a meaningful AR interaction?

What if someone can interact with each desk or table or piece of furniture? Someone's bodily movements around a classroom, building, school, or city may help them develop levels of understanding and meaning that are not possible from a website or textbook alone!

With AR, teachers can create richer experiences by making the most of embodied experiences!

10

Virtual Reality in the Classroom

Could new advances be useful for education?

Virtual reality (VR) has been dominating headlines lately with the development of flashy new gizmos that promise to help users see the world in new ways. However, VR is not a new technology by any means: it seems like it's been around for an eternity in technology time. Each iteration of VR headsets since the 1980s has used similar visual principles and goals over the last two decades. There are many newly released VR gadgets on everyone's gift lists. The suddenness of these devices going to market demonstrate that companies are committed to making VR cool again, so what can we expect to see this time around? More importantly, how can we teach with new VR and how do we get started with VR for education?

When we hear about reality in tech news, there are a few different technologies that are usually being mentioned. VR, and its cousin AR, both show users a digital reality, but these flavors of reality depend on how much of the "real world" is used to alter one's reality.

Thanks to VR, we can experience many new places and have Multisensory adventures inside the classroom. But has VR advanced enough for it to be useful?
Photo Credit: Knight Center for Journalism (via Flickr-CC[1])

Virtual reality (VR) is different from **augmented reality** (AR) in that AR simply "superimposes" digital information on the real-world things that we see. It does not try to create a completely new experience, but instead adds items to the existing surfaces of the space in which a participant occupies. AR is usually done through a tablet or smartphone today, but can also now be done with goggles (like with **Microsoft HoloLens**[2]). With AR, the canvas on which the technology animates objects is still the real physical world as no new spaces are being created.

On the other hand, VR creates an entirely new space: the physical is completely obscured from your line of sight. Users only see the new world that is created by wearing something that prevents you from seeing any physical things around you. When a user dons a set of VR goggles, they can be taken to a fantasy land, visit somewhere halfway across the world, or build their own world based on their interests.

---

[1] Knight Center for Journalism (2015). Virtual reality demonstrations. Flickr Photo. https://www.flickr.com/photos/utknightcenter/17191398541/
[2] Hardware- Microsoft Hololens: https://www.microsoft.com/microsoft-hololens

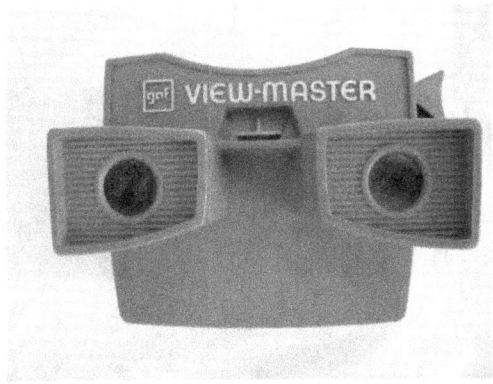

The ViewMaster may be old, but it uses the same principles of today's VR: block out the rest of the world to create a new one.
Photo Credit: Deiby Chico (via Flickr-CC[3])

VR today doesn't have to be animated, either. I tend to think of classic, 1980s VR as an animated world with blocky objects and angular landscapes. With today's VR, the resolution has increased significantly. You can visit exotic lands that are captured with a camera, take tours of historic places, or experience almost-real computer-generated vistas. The best feature, though, is the ability to turn your head and experience scenes in 360 degrees. Built in gyroscopes and accelerometers capture when you turn or tilt your head. As such, the screen's display moves with your head, as if you are standing at the scene and looking around.

Much like the ViewMaster reels of the mid-20th Century, you can see spectacular sights right from the classroom

However, you still need some specific equipment to experience full-range VR. The idea is to provide an experience that fully captures the line of vision. In 2016, the main idea is the same as it is a set of goggles that blocks out all light. The newest goggles are being

---

[3] Chico, D. (2008). ViewMaster. Flickr Photo.
https://www.flickr.com/photos/deiby/2741030309/

A Google Cardboard kit—nothing more than a piece of folded cardboard and a couple lenses. Drop your phone in and go on a VR safari. But don't let its paper walls fool you, it's a VR powerhouse!

produced by just a few companies, such as the **Oculus Rift**[4], **Samsung Gear VR**[5], or **HTC Vive**[6]. The challenge for educators with these headsets, though, is that they typically require high-performance computers in order to process the graphics that will be displayed in the goggles. The need for $500+ for a set of goggles and a $1000+ computer to support each set of goggles essentially eliminates the classroom practicality of VR, at least until the price comes down significantly.

There is a (very) low-cost VR alternative popping up in schools, though. Google had a crazy idea a couple years ago to make a device in which people can slip their smart device and wear over their eyes. It may look silly, but it's pretty effective at delivering full-range VR scenes. It's called **Google Cardboard**[7], which is a piece of folded up cardboard with a couple plastic lenses in which would-be VR enjoyers slip in their phone with a special VR app running. There are now hundreds of VR apps and 360-degree videos to enjoy for free in the app stores for both Android and Apple. The low-cost entry for this type of VR has been the most exciting,

---

[4] Hardware- Oculus VR: https://www.oculus.com/
[5] Hardware- Samsung Galaxy Gear VR: http://www.samsung.com/global/galaxy/gear-vr/
[6] Hardware- HTC Vive: https://www.htcvive.com/us/
[7] Hardware- Google Cardboard: https://vr.google.com/cardboard/

with Google even offering millions of cardboard kits to schools free of charge. After all, it's just a piece of cardboard!

In fact, our old friend The **ViewMaster**[8] is even making a comeback with their own cheap VR starter kit that uses a smartphone!

Thinking pedagogically about virtual reality

As I like to do with most techs I write about, I find it helpful to think about the goals that we would want to accomplish as educators with VR. It's important to think about the goals alongside what kinds of activities that VR can help people do to make sure the technology choices don't completely drive the learning experience.

Authenticity. If the tech continues to get better (and cheaper), VR offers us many classroom activities that are beneficial to learning. I think the biggest benefit that we can see is that it gives students more opportunities to have authentic, genuine experiences that use a wider sensory range. It is perhaps the use of multiple senses and the addition of context that gives an experience the "genuine" or "authentic" vibe. VR can provide both context and increased sensory involvement, making it a prime candidate for fostering richer learning experiences than simply reading a book or even watching a video. VR can easily couple audio with a full visual range, helping you *almost* be right there in the action. We still can't touch what we see, but the VR makers are getting good at letting us stand there and observe.

Interactivity and open-ended inquiry. This ability to "be there" extends into both historic and current events. We can now experience lands far away or interact with historic locations in ways that capture our senses more than a textbook, image gallery, or video can. Although the art of storytelling is an important way to experience things (such as in a documentary or narrative text), interactivity is another important aspect to learning that VR can bring us. What we see is based on our movements and interactions. In some VR worlds, the designers give participants the ability to affect

---

[8] Hardware- ViewMaster: http://www.view-master.com/en-us

change in the world. This obviously translates well into gaming, for which experts see VR to influence the most in the next couple years. But, this also translates well into learning, in which simulation and an interactive, multisensory world can come together to give experiences in which students can practice inquiry, ask open-ended questions, and explore.

This has most recently been seen in the game **Minecraft** (see Chapter 8), which provides an open-ended world that is built just for exploration. Minecraft has also been making moves to enter the world of VR, with players able to interact with objects with body movement. It will be interesting to see future applications of learning if we can think of creative lessons, projects, and challenges on which we can build from the exploratory contexts of Minecraft and other similar virtual worlds.

Live, real-time interactions. VR also gives us many new options for communicating information. Interactions between teachers and students can extend into the virtual world, where class sessions, meetings, and digital field trips can occur in places where additional context may help the learning goals, such as visits to historic locations. Living up to its web-based potential, VR also gives us the ability to host live gatherings of people not in the same physical location, allowing everyone to meet and interact in the same virtual space. Thus, the class discussions may be enriched by allowing students to leave the confines of the classroom. And yes, teachers can still lecture, too.

See things you can't normally see. In terms of what to teach, there have been many disciplinary subjects that have found themselves in the VR spotlight. For years, advanced skill disciplines have used simulations and VR to help hone their abilities, such as applications in medical, pilot, and advanced engineering education. Surgeons can practice on virtual patients, and pilots can fly jets made of bytes instead of aluminum. The ability to see phenomena, places, and events in new ways are a strong benefit to many scientific areas, such as chemistry, physics, biology, and astronomy.

For instance, in chemistry and biology, the microscopic and subatomic scales can now be represented and seen in virtual worlds with the visualization of molecular structures, cellular makeup, or protein interactions. The benefits of seeing in VR are not just limited to science, however. Social studies learners can visit historic locations, watch and participate in reenactments of key historic events, and even watch live events unfold. The potential for live observation was most recently was done on a large scale by NBC for the 2016 Olympic Games[9], in which VR users could watch their favorite events as if they were there. A new initiative by Google called **Expeditions**[10] hosts virtual field trips to both exotic and commonplace locations around the world, or even off-world up to the International Space Station! There really are countless ways in which we can use VR to give us new perspectives on our world, and new applications are being developed every day!

One final thought on VR, though. Although the new technologies are fancy, they are expensive. It may be best to use the devices already in students' pockets. And if they don't have a phone, phones are getting much cheaper (and a bunch of phones would be certainly cheaper than just one set of VR goggles and a computer). This faux-VR via cardboard may not be the highest resolution or the best experience. However, it's certainly the biggest bang for your buck. Give Google Cardboard a try—pick up a cheap kit from Google or Amazon, download a couple apps, and imagine the ways in which your students can really get some exciting experiences related to what you're teaching. Also, try it out because it's just plain fun :)

Just adding a little context and letting students have real experiences is a key to learning. VR just changes the game a little bit by shifting our definitions on what counts as "real."

---

[9] App- NBC Olympics 2016 VR: http://www.nbcolympics.com/news/experience-rio-olympics-virtual-reality
[10] App- Google Expeditions: https://www.google.com/edu/expeditions/

11

Does This Book Have Pictures?

Some thoughts on digital visuals for getting ideas across effectively

"That book...it has pictures. Picture books are for kids!"
  - said every person who tried to sound smart, ever.

But...what they don't tell you is that everyone likes pictures in books and other modes of communication. Illustrations, photos, and pictures are one step closer to how we naturally experience the world through our five senses. It makes sense why visual cues are likely to spark thinking and and make communicating more understandable. It's timely to think more about visuals today as educators, as imagery has become an increasingly significant form of communication in the last five years with the proliferation of mobile devices, contemporary messaging apps, and new forms of digital media like virtual reality and simulation. As educators, we should keep our finger on the pulse of these trends in how people communicate and use technology in order to find the useful pedagogical nuggets

Everyone enjoys a good picture book – it's time to embrace it!
Photo Credit: Kids Books—Mike Thomas (via Flickr-CC[1])

embedded within. In this chapter, I check out some of the forms of visual communication beyond the photograph that can be employed by educators to create meaningful experiences for their students by bringing concepts to life through visualization.

Pictures in books and "serious learning" media have long been assigned to the juvenile world, with anything less than a great wall of text relegated to "not educational" status. If it's got illustrations (and they're not historical photos), it tends to find its way onto the kiddie shelf. Heady books full of dense prose have stood synonymous to learning for many decades of formal education in America. However, the image element is making a comeback as digital tools get easier to use and more powerful for generating useful visual elements for storytellers, scientists, and everyday people to communicate to others what they experience.

If anything, the research coming out of the digital literacies and learning sciences are showing that people tend to learn more when a variety of media are used in ways that match the desired experiences, skills, and types of concepts that are being studied. A

---

[1] Thomas, M. (2008). Kids Books. Flickr Photo.
https://www.flickr.com/photos/urbanworkbench/2589425038/

new wave of digital media that has dominated people-to-people colloquial communication over the last 20 years must have some kind of useful element to it if everyone's snapchatting, drawing notes, and taking photos all the time. If today's media are any indicator, the old adage that a picture is worth a thousand words may be more true than ever.

Beyond the photograph: Digital tools for thinking visually

Educators have a treasure trove of apps, websites, and software to help them and their students create engaging visual elements. If we take a step back and think of a "visual" as something beyond a photo, we actually gain a lot of opportunities with new media to describe and explain abstract concepts, stories, and experiences alike.

The most common method of visualization today is the photograph. Every pocket today tends to have a high-quality smartphone camera at the ready, so it could be fun to take advantage of the ubiquity of everyday technologies for pedagogy. Teachers can bring ideas into the classroom by taking photos of objects, scenes, and contexts that exemplify concepts that are being discussed in class. By discussing photos taken by someone, there's a realness factor that's added in which the photographer can share the story on how the image was composed and why. Student projects can also include photo capture and editing via mobile device or conventional camera. Photo editing is becoming more approachable to everyday people through mobile device photo editing apps (e.g., **Pixlr**[2], **Photoshop Express**[3]).

Don't forget—with a mobile phone, you can also take photos or scan all of the conventional visuals as well: drawings, paintings,

---

[2] App- Pixlr: http://pixlr.com/
[3] App- Photoshop Express:
   http://www.photoshop.com/products/photoshopexpress

sketches and even doodles in the margins of pages. In terms of traditional artwork, it's becoming easier to do it digitally now, too. Although graphic design is certainly not new, drawing creations directly into a digital device by hand has become easier than ever. With a stylus (or sometimes, finger) and drawing apps on the iPad and Microsoft Surface like **Pixelmator**, **Assembly**, and **Photoshop Sketch**, visuals generated by hand can be made directly into bytes, often with interesting and fresh looks. These new drawing apps give great resolution and high fidelity to the movements of the hand. It's worth trying out, especially if your students have devices of their own—you'd be surprised at the creations someone can make on just their phone!

Sometimes you can get more out of your photo than just the image itself, though. Annotation and quick editing functions can add layers of information to an image. Today, this is most commonly seen on chat apps and social media like Snapchat and Instagram, where people can draw notes on images with their fingers and superimpose small graphics or "stickers" on the surface of an image. This creates an entirely new visual, with many streams of meaningful info running through it. Of course, in Snapchat, messages are intended for specific viewers only. As such, they may contain inside jokes, secret languages, and other meanings that are useful only to a handful of people. However, the ability to annotate and do minor edits to images of all types is a signature

Speaking of minor edits, another common use of images with annotations are in memes and animated GIFs. **Memes** are static images that convey an idea or feeling, usually in a humorous way. **GIFs** are small digital images that can either be animated or single-frame, and are also typically used to convey an idea with humor. Once the image format of choice during the mid 1990s, these images have become quite popular again in recent years. Memes and GIFs typically reference pop culture items in response to a conversation to communicate a little extra meaning.

Animated GIFs contain text annotations that help convey an idea in a better way than the text can convey alone. A symbolic image (meme) or video clip of a cultural reference (animated GIF)

accompanies some text to communicate an idea. The age of the response GIF, or an animated GIF that is offered in response to a comment or to convey a feeling, has been fully manifested on social media accounts, twitter streams, and chat apps. The use of memes and GIFs merits its own essay someday about their pedagogical uses. Memes are easily created from a number of sites that allow you to upload an image and annotate some text (I like **Imgflip**[4]). GIFs are easily made on **GIPHY**[5], which can also be used to search existing GIFs in the database. Give it a try—some GIFs are pretty funny and seem to apply to situations much better than words can convey. Annotation in this form can be a powerful way to capture pop culture and use it in the classroom to understand concepts.

Another popular way to use images lately has been in a number of apps that create image abstractions (such as **Prisma**[7], **Vinci**[8], or filters on **Instagram**[9] posts). These apps put a new spin on the stuff you've seen before by showing you photos in a new light, color, or shape. Algorithms are used to either transform photos to simpler forms or to make a photo look like a certain style of art, like you get with a photo in the Prisma or Vinci apps. Software can also be used to alter colors in a way that invokes

---

[4] App- Imgflip Meme Generator: https://imgflip.com/memegenerator
[5] App- GIPHY GIF Maker: https://giphy.com/create/gifmaker
[6] App- GIPHY search: http://giphy.com/
[7] App- Prisma: https://itunes.apple.com/us/app/prisma-free-photo-editor-art/id1122649984?mt=8
[8] App- Vinci: http://vinci.camera/
[9] App- Instagram: http://www.instagram.com/

This is a meme. It's a meme about memes. How fitting!

conversation and thought, which is common in Snapchat or Instagram filters. There are currently many types of filtering and transformation apps that are downloadable from the app store, and have become increasingly popular with teens and college students. Check these filtering and abstraction apps out to give a fresh look at sometimes tried and boring images!

But, there's no rule that visuals have to be frozen on the screen! As we have seen with GIFs above, and with the popularity of YouTube and other internet video, there's no reason an image has to be static. Animation is one avenue that teachers can use to convey ideas with the motion picture. **Powtoon**[10] is an excellent, free service that lets you narrate an animated video to convey ideas. Powtoon lets its users choose the animated subjects, the actions they do, and any other graphics that might appear on screen. **Animaker**[11], **Biteable**[12], and **Moovly**[13] are all other examples of web-based software that takes the guesswork out of

---

[10] App- PowToon: https://www.powtoon.com/edu-home/
[11] App- Animaker: https://www.animaker.com/
[12] App- Biteable: https://www.biteable.com/
[13] App- Moovly: https://www.moovly.com/

Three shots of me in my office from the Prisma app. I like how it does the sketch look. There are color-based filters that make your photo look like it's been giving a paintbrush treatment, but I like how the two sketch-like ones of myself here turned out. Look at the quality of some of the brush strokes!

animation. In short, these companies make it so you don't have to learn graphics production or programming in order to get some quick illustrations in motion on the web.

Of course, as YouTube and **Vimeo**[14] have become easier than ever to use, we could also add a video component to communications. However, producing video is a topic for another day and another article.

One interesting note about video, though. You can bring video into your flat, text-based documents such as papers, posters, and even flyers or business cards. Videos can be embedded into flat text with the use of a **QR code**, allowing a reader with a mobile device to get more info about what you're talking about. Once scanned, QR codes can take readers to a video about the topic they're reading, to a document with more info, or to other media that would be interesting. I have put QR codes on many informative posters I've made over the last five years, with links taking readers to quick videos of me presenting additional information, additional graphics, or important data.

---

[14] App- Vimeo: https://www.vimeo.com/

A QR code. This one will take you to the Wikipedia page that describes QR codes. Try scanning it with your phone!

Try making a QR code by using a QR code generator for your next project and try linking to a YouTube video!

Finally, today's plug-and play digital visualization tools allow us to bring data from any number of sources and make it into an informative picture. Having students find and use data sources (even simple tables) and integrate them into a visualization can help make concepts really come alive and understandable. When you give data a shape through a visual, it creates a completely new way of seeing the phenomena you are studying. The tool I use frequently is **Infogr.am**[15], which doesn't require any programming.

Thinking pedagogically about using visuals

There are some fun pedagogical strategies with digital visuals that can be used to get students thinking and engaged in active learning. Many of these strategies can be used regardless of the subject being taught, with educators just needing to adapt the strategies to meet any particular subject study goals and desired work output of students.

Sharing your experience. Storytelling is one of the most powerful ways of conveying information. We are built to intuitively

---

[15] App- Infogr.am: http://www.infogr.am/

understand how things happen to people over time. Sharing experiences can only go so far if you use words only. This idea is certainly not original—people have been illustrating and creating visual art for centuries in order to pick up where words fail. However, with today's digital visual tools, students can experience the stories of others in vivid ways, as well as share their own stories so that others might learn from them. Giving space and time to concepts through concrete visual examples helps make ideas approachable. Try having students integrate a number of visual media into their next project with a specific direction to share their experiences and let the world see what they see. In addition, challenge students to take abstract concepts that are hard to put into words and to transform them into something visual (for example, feelings or adjectives). This exercise certainly borders on the arts, so don't be afraid to integrate elements or lesson plans from arts education as well!

Discovering the power of data. Students who use visual elements can quickly find the value of using data in their works. Teachers also can find new ways for students to experience data to make it approachable and interesting. For some students, the idea of math, numbers, and data tables can be intimidating. However, visualizing data can promote new ways of thinking about data and perhaps even break down some barriers to math and numeracy if they exist. Find ways to insert data visualizations about concepts that are being discussed in order to pave the road for further conversations with students on the value of data in storytelling, science, and argumentation.

One of these things is not like the others! Comparing and contrasting ideas and their various facets can be jumpstarted with visual media. Words have been the main workhorse of communicating ideas for centuries, but digital media gives us a bunch more tools for seeing phenomena. We can compare and contrast in many ways by looking for visual cues, as well as the abstract elements. Integrate some digital media into lesson plans, or have your students create digital media in order to give them opportunities for comparing and contrasting ideas. By putting visuals into the mix, you give students objects on which to focus their attention and analysis.

Working under constraints. Words can be cheap. Spark creative ways for communicating ideas and thinking by limiting the number of words that can be used to communicate an idea. Or, eliminate walls of text altogether and use visuals only. And this isn't just an exercise for a teacher putting together lesson materials. Having students frame their thoughts around visuals sans text might generate some new ideas or get students thinking. I find that when I put constraints around what I'm trying to do, I can get some interesting ideas.

Adding all the visuals. Compared to above, a similar experience can be had in the reverse: Having many or all of the visual elements described above in a collage, which again promotes deeper thinking into the meaning behind the inclusion of each piece of media. Loosely connected works that include a variety of media can be used to convey many facets of a complex concept in more ways that just words or a single medium can describe. From my experience, filmmakers of some of the best documentaries I've seen on a given topic include images, charts, and other visualizations in their discussions to further make the ideas they're describing understandable (disclaimer: at the time of writing this, I had just got done watching Ken Burns' Civil War[16] again, so it's fresh in my head). The benefits of visuals are numerous for visually communicating ideas.

---

[16] http://www.pbs.org/kenburns/civil-war/

12
Infographics to Infuse Data into Classroom Communications

Getting past "just words" to convey ideas

Using different ways to communicate concepts can help students better understand them: words alone may not always do the trick. By bringing photos, images, and videos into the classroom, you can help your students learn to communicate complex information visually, as well as help you as a teacher communicate complex ideas to your students using elements they understand. Plus, they tend to look way prettier than a table of numbers!

But how can we communicate data to students? Although multimedia are great ways to share concepts in multiple ways that help with differentiation, communicating about data and its various quantities and qualities can be difficult using just words. Lucky for us, some new tools have been gaining popularity over the last few years.

Infographics—they sit on the front page of the New York Times and are posted frequently on Facebook and Twitter feeds. They tend to help make complex data analysis and findings

102   Infographics to Infuse Data into Classroom Communications

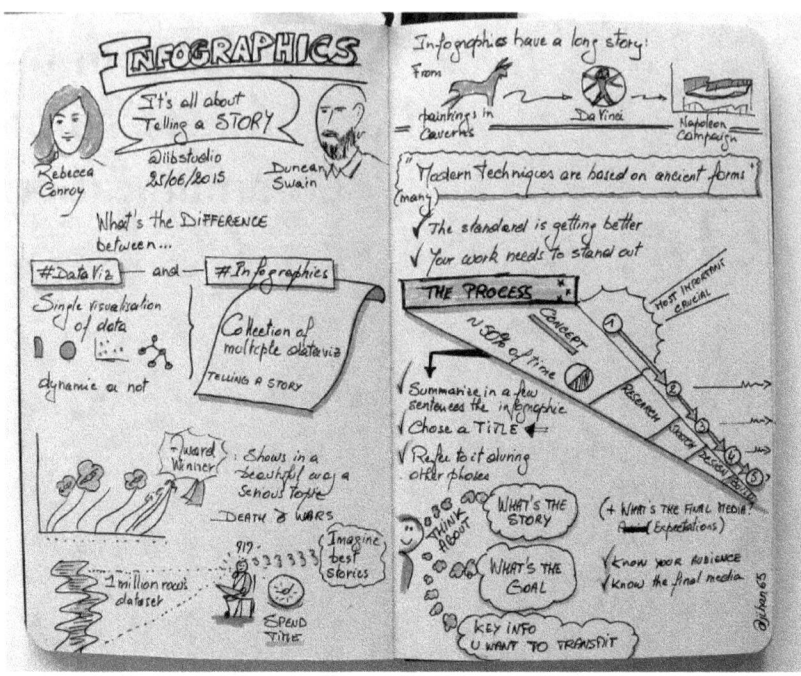

Telling stories with images is one of the benefits of new advancements in apps and technologies. Try some infographics out in your next lesson!
Photo credit: Infographic Sketches (via Flickr-CC[1])

approachable even to the youngest of readers.

There are lots of things out there that are considered infographics, so here's a short definition: **Infographics tell a whole story by using different pictures to both illustrate data and describe concepts visually.** Compared to graphs or other illustrations, the goal of an infographic is to tell a whole story—*the graphic can stand on its own*. I like to think of infographics as the merging of art and data, as some of the most beautiful illustrations I have seen are infographics.

---

[1] http://www.flickr.com/photos/27378944@N05/18557715533

No experience necessary!

It's true that infographic design requires some skills. Many of the infographics you see at news websites, blogs, and magazines are designed by graphic designers and require some programming. However, there are many new services via the web where you don't need any programming experience to build a nifty infographic!

Here are some of the infographic services that I've tried and like. These all have a very small learning curve and are as simple as entering or importing a spreadsheet of your data. They include many different ways to visualize your data, import other images, enter text, and compile a full story in the infographic.

Infogr.am[2]. Infogr.am is a free service that has many different graphs and visual elements from which you can choose. Assemble a full infographic story easily and download or embed your infographic into a website. In addition, many of Infogr.am's infographic elements are interactive, as readers can click on various parts and have them provide different information on-demand.

Google Chart Tools[3]. Google provides many free charting and graphing elements. Simply upload or hand-enter your data and get embeddable or downloadable images to use when building an infographic. The designs are somewhat more simple in their artistic quality, but powerful. Import your outputs from Google Chart Tools into MS PowerPoint, MS Word, or the free **Pixlr**[4] graphic design app to create a full infographic story.

Easel.ly[5]. Easel.ly provides a lot of free, high-quality design elements to help you make your data into stunning visuals. This service is particularly good at developing visuals for communicating relationships between people, places, and things. Many of the tools are aimed at drawing lines or highlighting relationships between these types of objects in an infographic.

You may also want to use more than one service to build an infographic, as there may be elements from different services that you like. With so many free services out there, you can find some great design elements to tell some stories. A quick google search for

---

[2] App- Infogr.am: https://infogr.am/
[3] App- Google Charts: https://developers.google.com/chart/
[4] App- Pixlr: https://pixlr.com/
[5] App- Easel.ly: http://www.easel.ly/

"make an infographic" will find you many more services not listed here, as well!

Going beyond pie charts: Compiling a story with visualized data

OK, so you have an idea of what you want to communicate. But how do you get started with making infographics? There are many free and easy to use infographic services out there, but how do you get started in telling a story with visualized data? There are three general steps that most people take when getting into the nuts and bolts of building an infographic.

Step 1—Find some data
The first step to making a good infographic is to find some compelling data. What kind of story are you trying to tell or what are you trying to teach with your infographic? If you have your students make infographics, what kinds of things do you want them to communicate? After you figure out the main points you want to make, you need some data. The hallmark of infographics is data What kinds of data would be helpful in making this point? Think of comparing different types of data or measures, or showing the relative size of something. You can even map out data geographically, showing relationships based on distance. Data can even be lists of concepts that have relationships between them and have different relevant importance between each other. You must have some data, however, to make an infographic. Otherwise, you are just drawing pictures for which there is no additional information being conveyed by the visual elements of the infographic.

Step 2—Draw
Once you have some data, you want to consider what to draw. Think of the key actors, variables, or concepts that you want to talk about in your infographic "story" and the relationships between them. Once you identify the "cast" of the infographic, it's time to figure out how to represent them via graphical elements.

There are many ways to convey descriptions, meaning, amount, and comparisons via infographics. I call these ways the "

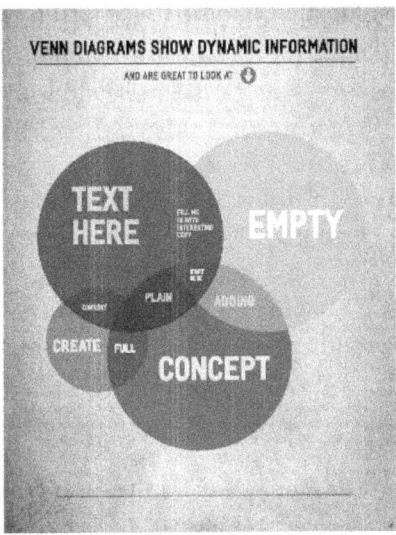

*There are many different dimensions that you can use
in an infographic, such as color, shape, size, and spacing.
Photo credit: easel.ly[6]*

dimensions" used in infographics, and each one can be used to compare, contrast, and show the relative importance of concepts you are measuring. Some common dimensions of infographics include:

- **Color.** Color is a great way to label or describe data. Different colors can be used to show membership in different categories or groups. Colors can also be used to show similarity between elements (such as similar but different colors, like different shades of green).

- **Shape and Icons.** Different shapes and icons can also be used to draw the attention of the reader and help make it easy to describe different types of data. Common examples of abstract icons are circles, squares, and stars. However, you can

---

[6] http://www.easel.ly/create?id=https://s3.amazonaws.com/easel.ly/all_themes/vhemes/venn/&key=pub

get creative with this, as different items can be represented by really anything: you get to decide how to illustrate your data! This is where the icons and shapes you choose can help describe the key elements. For example, icons can be the shape of flasks for infographics related to chemistry, wolves or trees in infographics describing wildlife, or cell phones in infographics describing the relationships of technology with people.

- **Connections.** Drawing lines between elements can signify relationships. Make good use of lines and connections in your infographics. This is particularly valuable when coupled with the "distance" dimension below!

- **Size.** The size of an element could be meaningful if you make it that way. Think of bigger and smaller circles—this provides a great way to compare the relative size or impact of data information.

- **Distance and Geography.** You can superimpose elements (such as bars, circles, etc.) over a map to show the relative distance between objects. Or, you can use a map to make the distance between objects meaningful in some way (such as the number of units on an X or Y axis).

Step 3—Get artistic!
The best infographics are those that synthesize storytelling elements, data visualizations, and artistic elements. This is where you or your students can really get creative! Not only are data important in the infographic, but so are other images, embellishments, backgrounds, and color schemes. All elements in an infographic help tell your story.

Some of the artistic elements you might want to include are headers, images of actual things, places, or people, images that help describe concepts, background patterns and colors, and the fonts

You can get artistic and add other storytelling elements to your infographic, such as the trees, people, and path in the infographic above. Photo credit: Walkway via easel.ly[7]

you choose. Many of the drag-and-drop services listed above include many stock elements that you can choose from. If you're adventurous, you can try your hand at creating some of your own elements and images using free graphic design software such as **Pixlr**[8] (similar to Photoshop).

Many of the drag-and-drop infographic services also provide some good examples of how to infuse art, text, and data to create a compelling story that keeps people's attention.

I think the best way to get inspiration for your own infographics, though, is to check out infographics when you see them in the news, on blogs, or on your Facebook/Twitter feeds. In particular, pay attention to how they use images to compare and contrast data and how they infuse artistic elements into the overall in-

---

[7] http://www.easel.ly/create?id=https://s3.amazonaws.com /easel.ly/all_themes/vhemes/walkway/&key=pub

[8] App- Pixlr: https://pixlr.com/

fographic. The best part of infographics is that no two need to look alike, unlike regular boring old bar charts. There have also been some wonderful compilations published over the last three years about infographics that appeared on the web or in news publications. Check out the most recent volume from 2015!

Part 2

Trending Topics and Stuff to Consider

13 What Makes Technology "Smart" in the Classroom, Anyway?

"Smart" technologies are everywhere, but what about in the classroom?

We hear it all the time: "smart" technologies are everywhere. Smart TVs, smart boards, smart desks, smart phones, smart cities, smart power meters, smart pens, even smart kitchen appliances. But what does "smart" really mean in relation to technology? These techs certainly don't have artificial intelligence, so what is it that makes them smart? And how can that influence learning in classrooms?

Technology's intelligence

With advances in internet-based technologies in recent years, it seems that every object we have in the house, workplace, and classroom is connected to the web in some way. In many ways, simple things like coffee cups and pencils are becoming computing "devices" on their own. Basically, this is being driven by the fact that more and more objects can collect data and interact with a computer. Data are being collected about much of how we live, from how many steps you take each day (via devices like **Fitbit**[1]), to how many miles you drive to calculate insurance rates based on how much you drive (via tech like **MetroMile**[2]), to automatically counting the calorie contents of your morning cup of coffee (via the **Vessyl cup**[3]). But what happens to all these data collected from these techs?

All of the technologies mentioned above are considered "smart" in some way. So, what technology features give our devices some smarts? As I was writing this article, I was able to think of five features off the top of my head that give some "intelligence" to new tech. If you think about technology in terms of these features, you'll notice a lot of these popping up in new classroom technologies.

Connected to other apps and hardware via the internet. I think the most pervasive characteristic of smart technologies is that they are connected to the web. In fact, they are not just connected to the web, but smart technologies are also connected to other devices and apps. The output from one app can become the input of another, making a long chain of connected apps that can help your life.

Emphasis on data. Almost every device that is "smart" collects and uses data in some way. In my opinion, that's the primary value of smart technologies. They capture information, which can be used later on by the device itself, other devices, or the person using them. Although there are some examples in which data or information aren't collected, try thinking of a "smart" app that doesn't collect info!

Interactive. With most smart technologies, you can usually *do something with them*. They are interactive to some degree. You, the person with the technology, can change settings, get information, or cause the technology to perform some task. Interactive elements can keep people engaged and interested in an activity.

Automation of tasks. Smart technologies often do their tasks automatically. For example, your **Fitbit**[1] automatically counts steps and your **Livescribe**[4] smart pen automatically captures notes into

---

[1] Hardware- Fitbit: https://www.fitbit.com/
[2] Hardware- Metromile: https://www.metromile.com/
[3] Hardware- Vessyl: https://www.myvessyl.com/
[4] Hardware- LiveScribe: https://www.livescribe.com/en-us/

your **Evernote**[5] account. You don't have to do many of the tasks in between. You automatically get your notes into Evernote in multiple formats instead of having to transfer the data out of the pen into the Evernote account. The new **Amazon Echo**[6] or **Amazon Dash Button**[7] smart home devices can automatically order groceries or home goods when you are running low. Smart technologies tend to remove some of the steps between getting stuff done.

Decision-making possibilities. This is really going to take off in the next few years I think. As I mentioned data collection and automation of tasks functions above, smart technologies often make decisions based on the data they collect. When they reach a certain state, or they collect data that meets certain criteria, smart devices can make decisions on their own. A lot of these things happen in the background. For example, when you hit your 10,000 steps on your Fitbit for the day, it automatically makes decisions to give you the star for the day and reward you with a congrats message. Likewise, when you finish a presentation on a smart board, it will make a decision to automatically save your presentation and annotations so they are not lost.

All these devices connected to the Internet can seem scary. This scenario can get pretty dystopian pretty fast, but it doesn't have to be. it can also be beneficial to us if we use these techs in a smart way, especially in education. This is where the "smart" part of technology comes into play. Technologies that can provide us with interactive options, help us make decisions, and make decisions on their own by examining data based on their programming.

The bottom line about smart technologies is that the tech is only as smart as its programming or the rules that have been set. Technologies with limited purposes to help you perform certain tasks in the classroom should be sought out, while erroneous or unnecessary technologies should be abandoned. Every smart tech you choose should be helping you accomplish actual pedagogical activities—you shouldn't be bending your goals to meet the needs of the tech.

I like to say that computers are *really smart, dumb machines*. They only do exactly what you tell them, but they do it well. A

---

[5] App- Evernote: https://evernote.com/
[6] Hardware- Amazon Echo: http://www.amazon.com/echo
[7] Hardware- Amazon Dash Button: https://www.amazon.com/oc/dash-button

principled use of smart tech would be to identify what technologies help you do and adapt them to meet your needs (not the other way around). Find out what the smart elements of technologies are and what activities they help you accomplish to get a handle on how you can use smart tech to your advantage.

Thinking pedagogically about smart technologies

So, we have a lot of smart technologies out there, but what should we do with them? Aside from smart techs being more inherently interactive and thus more engaging, there are some other key pedagogical features that smart tech can bring to a classroom. It's usually helpful for me to approach new tech in terms of what kinds of activities it can help me do rather than think about what a technology does. For this chapter, I've brainstormed a little list below on some of the things you can do with smart tech in support of the pedagogies you use.

Setting expectations and constraints in the classroom. Smart tech can be used to set constraints, expectations, and rules in the classroom. Technology is built on rules, so it is well suited to help set constraints in your classroom. With smart technologies' abilities to collect data and make decisions, you can use smart devices as indicators for when and how to act. In addition, smart technologies can turn on or off their interactive capabilities based on rules you program, and can be set to turn interactive functions on and off based on received data or inputs. Smart tech can set the constraints for activity. Use the constraints and "rules" of technologies to your advantage!

Offloading some of your work. There's a nifty service out there called **IFTTT**[8], which is short for "If This, Then That." This service lets you link up some of your most-used apps and have them talk to each other. After creating a few of these links (called "recipes" or "applets"), you can create some automated tasks that are done by the computer whenever you do certain activities[9]. As a result, a lot

---

[8] App- IFTTT: http://ifttt.com/
[9] IFTTT keeps a list of sample popular recipes or "applets" on their site for you to try out: https://ifttt.com/recipes

of your work can be offloaded to smart apps, saving you time and headaches. In effect, you can create your own personal assistant to help you with some of the smaller, day-to-day tasks you have to do. I find it helpful to offload some of the mindless tasks (such as checking the weather) that I don't always remember to do. There are some great samples of recipes you can use to help make your life easier, and some helpful blog posts out there on educational applications of IFTTT to help automate some tasks and offload work[10].

Coaching students. Smart techs can serve as "coaches" for students as well. With data playing a prominent role in smart tech, once certain criteria are met (or not met), notifications, suggestions, or congratulatory notes (e.g., badges) can be sent to teachers and students. Take advantage of the data collecting and decision-making capabilities of smart tech to have devices and apps serve a coaching role in your classroom!

Directing traffic by being a referee. Smart tech can also direct traffic in the form of a referee. Some tech can be used to be the middleman in group work or facilitate activities. Some techs can be set to tell groups when they should be working, while others can be used to display information to students and teachers to help regulate activity or conduct formative assessment. If data is being collected, use it to help guide the workflow of students!

Also, a word of caution. I've mentioned it in other chapters in this book, but it's important to seriously consider what data are be-

---

[10] Here are just a few URLs to blog posts that I found when researching this chapter that talk about IFTTT applets for organizing educational work:
(a) Williams, G. (2011). How to use IFTTT (And why you might want to). *Chronicle of Higher Education*. ProfHacker Tech Blog. 14 September, 2011. http://chronicle.com/blogs/profhacker/how-to-use-ifttt-and-why-you-might-want-to/35973
(b) Wilson, M. (n.d.). If This Then That – IFTTT in the classroom. *ICT for Teaching and Learning in Falkirk Primary Schools*. https://blogs.glowscotland.org.uk/fa/ICTFalkirkPrimaries/if-this-then-that-ifttt-in-the-classroom/
(c) Skrabut, S. (2013). IFTTT Learning Guide. Wiki website of University of Wyoming Extension http://www.wyomingextension.org/wiki/index.php5?title=Learning_Guide:IFTTT
(d) Houston, T.J. (2013). 5 #IFTTT recipes that make my life easier. Personal Blog. 30 January, 2013. http://tjhouston.com/2013/01/5-ifttt-recipes-that-make-my-life-easier/

ing collected by tech providers and how the data are being used. This is especially important if you are using tech with minors in K-12. Special care should be taken to care for the privacy and data of students when they interact with smart technologies. It's worth examining the data practices of each smart technology provider you choose because smart devices are so dependent on data.

14

Thinking about the "Making" Movement for Education

Promising creative outlets for getting ideas out of heads and into real life

Making has become a popular hobby for people of all ages. It's inspired many new companies to create various digital kits, programmable computers, and cheap 3D printing devices to help people get ideas out of their heads. There's even a magazine[1] dedicated to the unique projects people make. The educational implications for making seems obvious on the face, but what do we really get pedagogically from making when it is coupled with classroom learning? We find that making can really immerse students in custom, personalized projects, but can also be a point of difficulty without enough direction. Let's take a look at making in the classroom.

---

[1] Service- Make Magazine: http://makezine.com/

You can make all kinds of things in a makerspace. Physical prototyping tools give learners the opportunity to make ideas and concepts real by giving them physical substance-and with electronics and programming-the ideas can come to life. Photo credit: WeMake Milano (via Flickr-CC[2])

The value of making something

There's been a boom in simple, manipulatable devices and materials to help people realize their design goals. The cost of producing physical, interactive objects has significantly fallen in recent years. In addition, the learning curve on developing physical objects from scratch has also flatlined. In short, it's easier than ever to make something with your own hands from just an idea. Students who interact with makerspaces in learning can not only create custom physical objects, but can also make them interactive with digital tools and give them life. In effect, students' creations interact with them, helping represent concepts being taught. Students also learn multiple valuable lessons during the creation of their objects, as well.

So, why make something in the classroom?

First, I think the **physicality** itself that people can bring to their projects is an important aspect of why making is attractive for

---

[2] https://www.flickr.com/photos/wemake_cc/26219477231

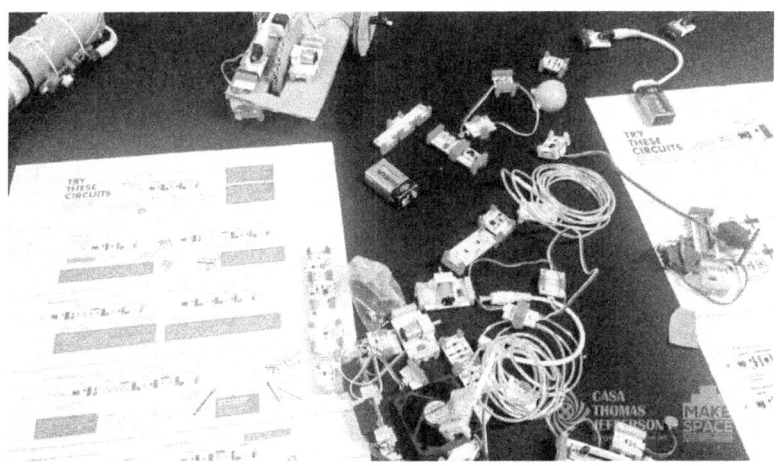

Robotics and electronic components can give life to concepts in ways that words can fail. "Making" adds a physical component to learning experiences that enrichens the educational process. Photo credit: Casa Thomas Jefferson (via Flickr-CC[3])

classroom learning. Creating a physical object and having it interact with the world makes it real when compared to something on a screen or represented by words in a text. Physical, embodied objects can engage all the senses and promote students' critical thought about why objects are the way they are and why they interact the way they do. Objects become non-ignorable conversation pieces in which students can imbue meaning and share with others the rationale for the design.

Second, it's the modern **digital interactivity** that really brings things up a notch. Making has been around forever: arts and crafts have been around forever, but it's entering a new revolution with real opportunities for impacting learning in all subjects. In today's makerspaces, it's not just physical objects that people tend to make, but instead interactive physical objects. Robots. Clothing that responds to sound or light. Boxes that open or close based on the time of day or a voice command. Artwork that shifts based on a mathematical formula, or a button, or someone's tweets. When you begin to *integrate input and output rules* and *sensors* into physical devices, the possibilities begin to seem endless. These interactive ob-

---

[3] https://www.flickr.com/photos/ctjonline/26257711515

jects are not just static—**they do something**, and people can do things with them. These create memorable, unique experiences that capture the imagination of students. Tied with learning goals, interactive physical objects can really help students represent ideas, concepts, and stories while giving them complete ownership and creative control over their project.

Third, I think that with physical prototyping today, **creativity is boundless**. There are just so many combinations you can try between sensors, inputs, outputs, and devices when trying to make creations interactives. In addition, when considering devices that create physical objects such as 3D printers, milling machines, and engravers, quite literally any form can be given physicality. You can create multiple spheres, or you can create a snowman from spheres. Simple boxes to sci-fi robots can all be designed in computer-aided design (CAD) programs and turned into physical forms. Ideas can become physical realities. Creativity and customization has become the norm in classrooms that embrace making.

Finally, I would say that the **complexity** involved in projects is enough to keep people really engaged. There's no straightforward answer to a problem when you are creating a custom project. You have to balance a lot of variables and ideas to get your ideas to work. This can be really gratifying to do and helps students gain mastery over the concepts at play. However, it can be a double-edged sword. Because there are so many factors at play in any given project, any one of them can cause a project to come to a grinding halt. Not knowing how to connect electrical components (and not being able to find help on how to do it) can cause extreme frustration and ultimately, lead to giving up on the project. Part of the value of making is that it encourages long-term commitment and not giving up on a project. However, with so many elements at play, it becomes important for teachers to consider the balance between students getting help on a project, having someone do it for them, and having them struggle a bit for the sake of learning. I can say from personal experience that the struggle is indeed frustrating, but is really rewarding once you solve the problem.

The M3D printer makes quality prints and is priced around $300, making it an attractive choice for school and classroom makerspaces!

Makerspaces popping up in schools and classrooms

A buzz term in education talk circles is the makerspace, or a dedicated space for students to create objects. **Makerspaces** tend to include equipment for developing physical objects from digital files, such as 3D printers, milling machines, and plastic molding machines. They also include a variety of parts, devices, and kits for integrating digitally interactive elements in their projects, such as **programmable microcontroller computers** (e.g., **Arduino**[4], **PIC**[5], **Raspberry Pi**[6]) to handle rules on input and output, **conductive inks** and **conductive fabrics** to integrate digital components onto surfaces, clothing, and other objects, electronic components and sensors of all kinds to provide enough raw materials for students to develop custom input and output capabilities, and dedicated staff to help students conduct research, guide them on solving design problems, and to help them when they are stuck.

Makerspaces are not limited to just school-based facilities, though. Some classrooms are becoming mini-makerspaces. Some

---

[4] Hardware- Arduino: https://www.arduino.cc/
[5] Wikipedia (n.d.) PIC Microcontrooler. Wiki Page.
https://en.wikipedia.org/wiki/PIC_microcontroller
[6] Hardware- Raspberri Pi: https://www.raspberrypi.org/

teachers are getting small digital kits for making physical objects interactive. Some classrooms are getting cheap 3D printers of their own, such as the **M3D Printer**[7] to help students visualize their ideas in real space. If you want to think about making from a broad sense, even basic arts-and-crafts projects with cardboard and glue are considered making. However, making really takes off when you integrate the digitally interactive component to projects. It's cheap enough now that every classroom could have Arduino kits, electrical components, and a 3D printer in the foreseeable future. As these possibilities get cheaper, it will become more common to see the makerspace being everywhere, at every student's table, and not just limited to a specific place.

If you don't have access to maker tools, don't fret! Many libraries offer access to such tools. Additionally, cheap but high-quality 3D printers (such as the **M3D**[7]) are increasingly available that can transform digital drawings into physical objects by "printing" your design with plastic ink that dries hard. Programmable **Arduino**[4] or **LittleBits**[8] kits are cheaper than ever, allowing students to program rules for sensors, buttons, and internet input into their designs. You can even be a maker by building ideas with LEGOs, either with regular bricks or their programmable **Mindstorms**™ kits[9]! There's even **Make Magazine**[10], which provides inspiration and how-to tutorials on how to accomplish many things. Online providers like **Maker Shed**[11] and **Sparkfun**[12] provide ready-to-make kits and sensors for projects. And, to top it all off, YouTube and other "maker" websites have a treasure trove of online tutorials on how to do realistically anything digital for making.

If you don't have access to a makerspace, consider starting to form your own. All it takes is the will to start collecting resources, both physical and those within your network. Yes, things cost a bit of money to start out, but it's cheap enough now that the barriers to entry are lower than ever.

The key challenge with any makerspace, though, is that it needs to be used by educators in a principled way. By coupling education-

---

[7] Hardware- M3D Printer: https://printm3d.com/
[8] Hardware- LittleBits: http://littlebits.cc/
[9] Hardware- LEGO Mindstorms: https://www.lego.com/en-us/mindstorms
[10] Service- Make Magazine: http://makezine.com/
[11] Service- MakerShed: http://www.makershed.com/
[12] Service- SparkFun: https://www.sparkfun.com/

al goals with makerspaces, teachers need to know what goals they hope their students will get out of using the space for a class project. Just throwing students in a makerspace and telling them to learn is not great for achieving specific learning goals. In fact, without enough parameters, goals, or a specific project idea, students may be paralyzed in choosing what to do in the makerspace. Without a specific goal in mind, the sheer choice of machines, tools, and kits could make coming up with something to do really difficult. Sure, students will still learn by playing around without any guidance. The novelty of all the maker tools will be enough to hook the interest of many kids, and ideas can flow from there. However, to achieve learning goals that are attached to a specific class, students need to have an idea before they approach a makerspace to realize these learning goals.

15

Learning To Trust the Crowd

The quality and educational value of Wikipedia and other editable, crowd-based information sources

By now, any person who uses the web every day has heard the arguments against Wikipedia, the web's favorite collection of information. With anyone being able to post information to a page on Wikipedia, how can we be sure that information is accurate? Can't anyone just go in and trash a page for no reason other than to mislead people or to push their own personal agenda or bias? It's true that these are risks with using a crowd-based system like Wikipedia, but a closer inspection shows that many of these risks are mitigated by a closely-knit community of knowledge creators and curators. There are also lists of longstanding rules, procedures, and projects that determine protocols for decorum and editorial power. Something this organized that persists to improve quality can't be completely bad, right?

As such, there are a few reasons why we should trust the crowd when it comes to using and learning from information on the web. In fact, crowd-based systems like Wikipedia offer many pedagogical insights on how information is organized on the system and how

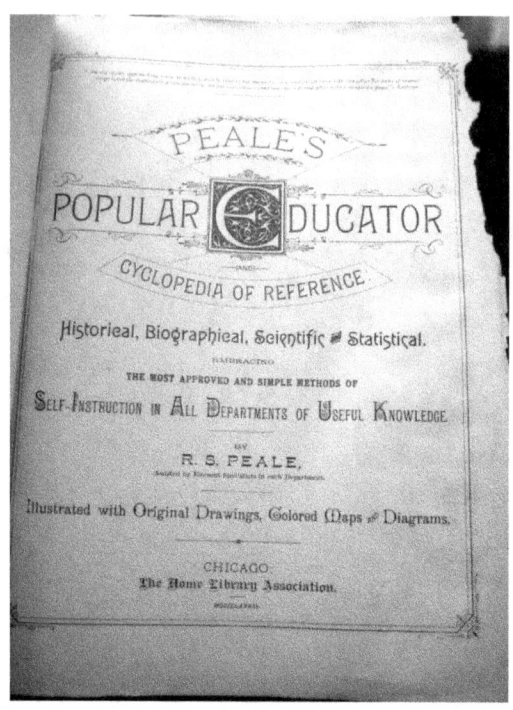

Are our ideas about information authority changing? In many ways, yes - just look to Wikipedia's 15 years of history. We're finding valuable sources of information where information can change rapidly to stay current and be trusted to be accurate, even when authors aren't credentialed "experts." Photo credit: Bryan Peters via Flickr[1]

communities work together to curate and create knowledge. In this chapter, I continue some of the thoughts from my article last week on digital information quality and tackle some questions about Wikipedia: the oft-cited example of both high and low quality, depending on which camp you're in. I take a look at some of the lessons for educators from Wikipedia and make a few suggestions on why trusting the crowd might be beneficial for learners in the 21st century.

---

[1] https://www.flickr.com/photos/urbandude/132020981/

The arguments for and against crowdsourcing

The ideas of **crowdsourcing** or **crowd collaboration** are not new: the more people that work on a project, the faster the work and the more work that can be done. *Many hands make light work.* However, in the Wikipedia sense, this old saying doesn't simply imply that the "crowd" of participants might just decrease the amount of work. Instead, it suggests that many hands might actually result in better work than any one expert or even a small team of experts are capable of producing.

As Web 2.0 technologies emerged in the late 1990s and early 2000s, these tools enabled many people to work together in real time regardless of geographic distance or having been formally introduced to each other in "the real world." Crowd-based work is best known by a large, open-ended number of people contributing to a project. However, the paradigm's other signature characteristic of *working on a real project with real implications* is often overlooked when attempting to design for crowd-based systems. People have to have something meaningful to do in order to show up to be a part of the crowd. It is this characteristic that shows us the value of integrating crowd and public projects into educational activities.

Wikipedia was launched in 2001[2] as a platform for sharing information widely and freely among people. Much of the world's information was locked away in libraries, books, and brains, with no fast way to retrieve it when needed on a moment's notice. The goal of building a repository of this knowledge that was both high quality and didn't include too much information as to stun a reader was a tricky proposition. It would take a lot of effort to get experts in their fields to write articles and keep them updated. The wiki concept promoted group sharing of information under the assumption that any one person could not possibly know everything, but a group of people could piece together information more efficiently and effectively. Anyone could participate in the collection, curation, and distribution of information if they wanted to—but only as long as certain quality measures were met. In fact, the threat of sabotage and false information has always been an issue with any information source. From the earliest days at Wikipedia, any page that

---

[2] Wikipedia (n.d.). History of Wikipedia. Wiki Article.
https://en.wikipedia.org/wiki/History_of_Wikipedia

did not meet the quality measures (for instance, rules pertaining to organization, content, and opinion) could be removed. It was up to the community to decide, but not just one person or even a small coalition could make these changes unilaterally: the community had to deliberate on the changes.

A strong principle of *deliberation* is implied in the Wikipedia model, in that any change should be discussed among community members. Because anyone can participate, the design philosophies of crowdsourcing and crowd collaboration rely on these rules and procedures that are agreed upon by the communities that use them. Without them, the platforms could quickly turn into chaos. The communities that voluntarily maintain Wikipedia pages are experts in their field, dedicated contributors, researchers, and people who have a general interest in topics. Expertise and authority are not ignored in Wikipedia articles, but in instead amplified as experts work with others to produce the highest quality page they can.

So, philosophically, crowdsourcing information represents a tension between one person's expertise and ideas, or what the "crowd" agrees upon to be necessary for publication. The community, through its deliberation, can in effect define what is important, not important, or even what is considered a fact or opinion. This, of course, is concerning, but is not new. Even with experts, the threat of inclusion or exclusion of information is an always present aspect of published information. Only with the crowd, we get a bit of transparency and can see how the crowd came to their conclusions of what was deemed important enough to include in an article. With an expert, we may never know what they were thinking or even what information we are missing in their publication. We only get the final product.

The question on the merits of who should judge quality of information today teeters between these two main perspectives. Both of these perspectives have real implications for internet-generation students who are learning to make sense of the web's flood of info. This debate becomes particularly important when we defer our evaluation to the judgement of others, which happens quite often in the age of needing quick information. We don't have time or resources to curate the world's knowledge for ourselves, nor judge the quality completely on our own, so we largely rely on someone else most of the time. Although we ultimately make a final judgement on whether we will use information, we often delegate either to *gatekeepers* or the *crowd*.

I tend to lean on the side of the crowd in terms of an overarching philosophy. I think more hands working at something is typically better as there are more eyes on the work and the processes by which decisions are made are transparent. Also, it's less responsibility on a single gatekeeper, even if they like playing gatekeeper. Information is only good if it can be used. It doesn't do much good behind a firewall, a library's doors, or in a human brain unless it can be applied to something. Power brokerage has always played a role in information distribution, as those who have more information typically can make better decisions (although this is not always the case). As such, the idea of a gatekeeper can actually be stultifying to improvement in the general sense, even if it's working to improve the quality of work. We're just at a point now where information can be traded faster than ever before. For the first time, it's feasible for many eyes to see it and provide feedback with rapidity.

The "democratization of information" argument holds that anyone is entitled to an opinion, and the Internet gives voice to these opinions. However, the loudest should not win, and certainly bullying your way to winning should not be allowed. Some people may find it valuable to hear the stories of individuals, and may find it high quality based on their particular needs for quality, so the crowd-based review approach can be really useful. It's also much faster and more efficient to get stuff done with a whole bunch of people than just one or a small team. Expert review is great in the case of a gatekeeper, but those same experts can participate in the crowd-based methods to share their skill.

However, as I reflected on the definitions I provided above, quality is based on the need of the person seeking information. As such, a person's information needs might be better suited with either one of the two paradigms of trust. There's no right answer here; each approach has its merits for various goals. I believe it ultimately comes down to what the information will be used for and the degree of trust we place with the information source.

Educators shouldn't discount information published by non-authority sources just because it is a collaborative effort of amateurs or novices. Instead, take crowd-based projects for what they are: people organizing to meet a particular goal. Keeping this in mind can help open up new value for Wikipedia and other crowd projects.

The community backbone of Wikipedia: Supporting curation and collaboration

Wikipedia is a useful case for examining how the community-based system might at times be just as good or better than a conventional gatekeeper, such as a news agency or academic article. By looking at Wikipedia with a technologist's eye, curious educators can uncover some unique features of the platform that demonstrate how curation and collaboration can go hand in hand.

Structurally, certain functions are built into the Wikipedia technology to support community collaboration and curation. The primary structure is a robust system for communication among registered "Wikipedians," who are community members that serve as the volunteer editors of each page. You may not think about it when first visiting a Wiki page, but each of the thousands of articles on the site has a "talk" page on which both novice and veteran editors discuss changes. Next time you're on Wikipedia, just for fun, try clicking the talk page tab at the top of the screen to see the in-depth discussion that occurs about changing the main front "article" page. Talk pages reveal that what you see on a main Wikipedia page is only the tip of the iceberg. Most of the work happens behind the curtain…only the curtain is transparent and allows all visitors to see the moving machinery, workers, and the messy discussion that leads to the distribution of knowledge.

I haven't yet seen a good answer or model as to why a volunteer chooses to participate on Wikipedia. Maybe they're passionate about a few topics that have pages and they want to contribute. Maybe they find zen in making edits to grammar and punctuation. Maybe participants want to feel like they are working on something that will last. Maybe participants simply want to be a part of a community and it gives a sense of belonging. In 2012, I found that this sense of camaraderie may be reason enough, as I attended the annual Wikimania conference[3] that happened to be in the town in

---

[3] https://meta.wikimedia.org/wiki/Wikimania

which I was living. Thousands of people love being Wikipedians enough to justify traveling from around the globe to discuss Wikipedia projects and its future. Maybe having something fun and meaningful to do is all that many desire.

I remember reading somewhere that 80% of Wikipedians are rule enforcers or what I call "quick editors"—the type that enjoy making immediate changes to Wikipedia pages that reflect current events. For instance, in the case of the recent Rio Olympics, medal counts were revised in real time by quick editors, who find enjoyment in making changes as they occur in the real world. In addition to rule enforcers and quick editors, the small remainder of participants are typically experts that contribute the deep knowledge that each page needs to stay relevant and correct. But, in any case, Wikipedia encourages any member of the community, including new members, to jump in and *be bold*[4]. This suggests that participation of any kind is the name of the game: you just gotta do it instead of talking about doing it!

Regardless of the motivation to participate, though, it is obvious that Wikipedia is fueled by its community. Each community member has a profile page and can participate on an unlimited number of articles for which they are passionate about. Articles would simply not exist if it wasn't for the volunteer effort. This curation community provides a living, documented example of how knowledge is proposed, transformed, and accepted, which can be a valuable laboratory for educators and students alike.

In addition to the formal communications channels at Wikipedia, there are a number of both written and unwritten rules that have emerged over Wikipedia's history as the community has formed. Membership required to edit a Wikipedia page. As such, it typically requires the approval of a page's community in order to make it to the front page permanently, with of course the word *permanently* being used loosely here. It's only permanent as far as the community decides it is *current* information and deserves to be on the front page. Of course, some changes by users don't go through the process, get through the cracks, and edits that weren't agreed upon get made to a page. However, depending on the speed and level of involvement of a page's community of editors, any *not*

---

[4] Wikipedia (n.d.). Wikipedia principle: Be Bold. Wiki Article. https://en.wikipedia.org/wiki/Wikipedia:Be_bold

agreed upon information will likely be removed as soon as it is spotted. Volunteer editors are also alerted to any changes, as a changelog is clearly visible to any participant. Although you can create an account that doesn't use your real name, all changes can be traced to specific users. Users can face repercussions for not following the rules, be it posting without talk page approval, or overtly trying to sabotage a page.

Speaking of rules, the above structures require rules in order to both be of high quality and to be self-governed. The rules are all decentralized so there's no one person in charge, and no one person can write new rules. Everyone accepts the rules and the rules are regularly debated. What makes it tough, though, is that some rules are written and some are unwritten. Unwritten rules must be learned through participation in the community.

Wikipedia editors are required to adhere to three complementary principles when contributing content to articles. First, articles must not be original research, in that they must not contain arguments, analysis, or insights that have not been elsewhere published. Second, to support the *principle of no original research*[5], articles must be *verifiable*[6], in that all statements made must have some source of attribution, even if not all statements are attributed to sources. Finally, all articles must adopt a *neutral point of view*[7]. Wikipedia's editors accept that multiple points of view exist for any given topic, and that no specific viewpoint or perspective should be promoted as the authoritative view on the topic. Despite its reputation in some circles, Wikipedia strives to provide high quality articles for every page, and they do a great job doing so. The three rules above help realize that vision, but each article's community of editors are the true heroes. Community members donate hours of work to ensure that all of the text and other media on articles adhere to these principles and communicate via article talk pages to discuss edits.

As a result, these are all excellent examples of how multiple disciplines work through discourse that requires participants to

---

[5] Wikipedia (n.d.). Wikipedia principle of no original research. Wiki article.
https://en.wikipedia.org/wiki/Wikipedia:No_original_research
[6] Wikipedia (n.d.). Wikipedia principle of verifiability. Wiki article.
https://en.wikipedia.org/wiki/Wikipedia:Verifiability
[7] Wikipedia (n.d.). Wikipedia principle of neutral point of view. Wiki article.
https://en.wikipedia.org/wiki/Wikipedia:Neutral_point_of_view

learn both the written and unwritten rules of participation. Wikipedia again provides an excellent laboratory and learning environment for identifying procedures, rules, and ways of working when participating in a real-world project are valuable skills for 21st century success.

Reconsidering Wikipedia for education: Pedagogical implications

Don't close the book on Wikipedia and keep it from the classroom—give it a chance to show its benefits to teachers and students of all ages. Of course, Wikipedia shouldn't be cited as a primary source in research papers. However, that shouldn't be the only thing holding educators back from using it for learning activities.

Educators can draw some interesting pedagogical insights from Wikipedia's structural design and community behaviors. There are some positive 21st century skills and learning outcomes that can occur when interacting with crowd-based information sources. We can also imagine some interesting teaching strategies and activities for students with an eye toward Wikipedia and communities like it.

Questioning how we use information. Wikipedia gives us a great tool for talking about how we get and use information. Do we just consume information and trust it inherently, or should we at least do a little thinking about the information we receive? It takes a bit longer and is tougher, but getting into the habit of asking at least a couple short questions about the ocean of digital information we use is an essential 21st century skill. Wikipedia's easy-to-read articles and fast access gives students ample opportunities to ask questions about quality and the decisions that were made by the authors, as each article is co-written by tens, if not hundreds of people!

Thinking about authority. Wikipedia's co-authorship style brings a new element into thinking about editorial involvement and how we place authority on sources. In some philosophical and linguistic circles, Wikipedia is also an excellent case for how social groups refine and define what knowledge is, what facts are, and how evidence is accepted in arguments (e.g., questions about ontology and epistemology). Being aware that we inherently place authority on sources and people is useful when assessing the value of information found on the web for a particular project. Authoritative sources (such as edited works and work from specific people) have definite value, though, as we don't always want to have to spend time judging a source's quality. Authoritative sources can be balanced with crowdsourced projects like Wikipedia, and it's a good skill to practice this balance.

Participating in knowledge creation. Wikipedia is the prime example of the participatory information democracy. Crowdsourced information is a living example of knowledge creation and debate at its finest. There are significant learning benefits to actively creating work products that represent knowledge, such as writing. Wikipedia and wiki technologies in general give students some wonderful tools for creating knowledge works and writing. If I had one suggestion for educators, it's to try a wiki on your own! You don't have to edit Wikipedia itself to practice knowledge creation, although Wikipedians are pretty welcoming to newcomers who honestly want to help. Setting up a class or project wiki is just as productive for students and would give them a real product that they can be proud of. Just be sure to take a moment to watch and reflect on the discussion and edits unfold for some very interesting, real-world examples of knowledge creation!

Matching information to the project. As with anything that is based in the digital world, the benefits of Wikipedia are only extended as far as they are useful for an immediate project need. Wikipedia is of course fun for learning trivia, but its real power is the immediate "starting point" for information that it provides when working on a project. In fact, Wikipedia editors try to only put "just enough" information on an article as to not bog down a reader and to keep the article approachable. If educators use Wikipedia for this purpose, it can be a powerful tool for helping spark student inquiry. Also, links to other articles help learners visualize connections and access related information. However, this surface-level review is also Wikipedia's limitation. It is really only designed to be a starting point...nothing more. Articles don't purport to go deeper than the surface, and that may not match a project's needs. Wikipedia helps anyone learning about virtually anything get enough information on a topic to continue their questioning, but they often have to go find authoritative and original sources in order to satisfy the needs for deeper inquiry projects.

16 Unintended Outcomes of Technology Use

Lessons for educators from the Pokémon Go phenomenon of Summer 2016

What happens when classroom technology isn't used in the way you thought it would be? Or, alternatively, what happens when students use technology exactly the way you intended, but undesirable outcomes happened? These are two questions that every educator has asked at some point as they attempted to integrate new technologies into their classrooms. Using recent examples from the Pokémon Go phenomenon, this chapter tackles these two questions with an eye toward how to adapt technologies for learning when they aren't meeting our expectations.

Since its release in July 2016, the increasing popularity of the Pokémon Go game has provided some casual tech observers (including myself) with a good opportunity to watch the different ways in which people use apps at scale. Altogether, I'd say the players have met the designers' expectations on how the game should be played: people walk around to catch and evolve pokémon, players work together in social groups to challenge and defend gyms, and

The Pokémon phenomenon has opened many questions about how to react when app use doesn't go as intended
Image Credit: Iain, Pokemon Go! Road Rage (via Flickr-CC[1])

they log in to the app regularly to keep up with their pokémon training goals. However, any use of technology can always have unintended consequences when activities interact with the other things that people do in their lives. Even if we don't realize it, we are often doing more than one thing at a time. As we use technology to facilitate an activity, it often interacts with the other things we are doing. Thus, it is at the *intersection of activities* where sometimes things don't go as planned.

There are some interesting early examples about the unintended outcomes of technology use that have been emerging from Pokémon Go as people go about playing the game. These lessons can be particularly valuable for teachers and instructional designers when making decisions about technology for learning.

---

[1] https://www.flickr.com/photos/iainstars/28306952343

Pokémon Go: A tale of unanticipated effects from tech-based activity

Whenever a new technology is implemented, it's notoriously hard to predict exactly how it will be used (or if it will be used at all!). This is because each time something is used, it's a brand-new context: every time someone starts to work on a new project, there is a unique combination of people, goals, and constraints that will influence the behavior of participants. Similarities between circumstances can help tech observers and educators predict how it might be used and what behaviors will emerge, but we can never be exactly sure. So, the best course of action is to try to anticipate some of the expected behaviors and to continually adapt the function of tech to meet these goals.

I'm sure this is what has happened at Niantic Labs, the group that makes Pokémon Go. In the initial stages of app design, it is likely that they anticipated that they wanted people to use the app to walk around, get some exercise, and capture Pokémon in real-world streets. They probably also anticipated the social elements of the game, intending for people to meet up at real locations and work together (or challenge them to a duel, if on an opposing team). We can assume these are some of the desired behaviors, as they are explicitly discussed in the training materials within the app. Although people are doing what the Pokémon game makers intended, there have been some unanticipated behaviors and outcomes that have occurred. These are the result of playing the game in a public setting in which movement and navigation are required.

When we talk about the "challenges" associated with Pokémon Go, it's kind of a kudos to the designers of the game. Pokémon designers made their game too well: people who play aren't taking their eyes off their phones! As a result, players have been walking into dangerous situations[2] without looking: blindly crossing streets, walking off of ledges like lemmings, and even navigating directly into oncoming traffic to find a pokémon. The app is succeeding at keeping people's attention, but also sometimes to the detriment of

---

[2] Horton, H. (2016). Police urge people playing Pokémon GO to look before they cross roads. *The Telegraph.* 07 July, 2016.
http://www.telegraph.co.uk/technology/2016/07/07/police-urge-people-playing-pokmon-go-to-look-before-they-cross-r/

players' other activities and safety. It's in this case we see the interaction effects of competing activities. You must walk to play Pokémon Go, but it cannot be separated from the separate, but parallel activity of walking in a populated area with cars, other people, and objects that you can bonk into.

However, even staying aware of one's surroundings hasn't been enough for some players to stay safe and avoid conflict. Pokémon participants have been finding themselves in dangerous situations at no fault of their own, as criminals have been "luring" players[3] to public Pokémon locations and then mugging them at their arrival. Many of these crimes have been reported in the few short weeks the app has been online[4]. Challenges have also been found on car-filled streets, as players have been urged not to drive and play simultaneously. The driving issue has been a significant problem[5], enough so that Niantic instituted a driving warning in an update to the Pokémon app[6]. Almost comically, there's even been a report of a Pokémon Go player driving right into a police car![7] These unanticipated effects from regular gameplay are serious and sometimes random, but they only made themselves apparent after people started playing the game.

Additional aspects of play have been problematic from the perspective of non-player observers. Because Pokémon is a game that is played in public, it may also interfere with other social activities and rules in which the player is not participating. This has been the case particularly with places of solemnity and reflection, such as

---

[3] Yuhas, A. (2016). Pokémon Go: Armed robbers use mobile game to lure players into trap. *The Guardian*. 10 July, 2016.
https://www.theguardian.com/technology/2016/jul/10/pokemon-go-armed-robbers-dead-body

[4] Syracuse.com (2016). Is Pokemon Go dangerous? *Syracuse.com*
http://www.syracuse.com/us-news/index.ssf/2016/07/pokemon_go_dangerous_every_crime_accident_death_shooting_linked_to_game.html

[5] Seitz, D. (2016). A 'Pokemon Go' player accidentally killed a senior citizen while driving and gaming. *Uproxx.com*. 25 August, 2016.
http://uproxx.com/gaming/pokemon-go-driver-kills-woman/

[6] Mogg, T. (2016). Pokémon Go now wants you to confirm you're not driving while playing. *Digitaltrends.com*. 8 August, 2016.
http://www.digitaltrends.com/gaming/pokemon-go-driving-message/

[7] Bowerman, M. (2016). Driver slams into Baltimore cop car while playing Pokemon Go. *USA* Today. 20 July, 2016.
http://www.usatoday.com/story/news/nation-now/2016/07/20/driver-slams-into-baltimore-cop-car-while-playing-pokemon-go-accident/87333892/

places of worship, monuments, and protected areas. In the first week of play, the U.S. Arlington National Cemetery pleaded with players[8] to respect the cemetery and avoid catching pokémon on its grounds. That same week, pokémon players were chased out of the U.S. Holocaust Museum[9] for irreverently catching pokémon. In response around the launch of the game, lawmakers forwarded legislation that requires "pokémon stops" to be removed whenever requested and that some places remain untouched, such as nature preserves[10].

Many applications of educational technology will never have to worry about criminal activity or walking headlong into oncoming traffic. However, these problems that have revealed themselves after play are great lessons for how something innocuous like pokémon play can influence other activities. These *activity interactions* are what we should attend to as educators when we think about bringing technology into learning. These early Pokémon Go challenges make a pretty poignant case for fixing unintended outcomes when they occur.

Keeping an eye on tech use

Educators can take steps to monitor tech-supported activities and adapt tech strategies when undesired outcomes are observed. To start, it's important to make sure any tech use is principled. We should use tech for a purpose or activity and avoid letting the tech determine the activity. The following four actions can help educators maintain the goals of the activities in which their students

---

[8] Hersher, R. (2016). Holocaust Museum, Arlington National Cemetery plead: No Pokémon. *NPR, The Two-Way Blog.* 12 July, 2016.
http://www.npr.org/sections/thetwo-way/2016/07/12/485759308/holocaust-museum-arlington-national-cemetery-plead-no-pokemon

[9] Peterson, A. (2016). Holocaust Museum to visitors: Please stop catching Pokémon here. *The Washington Post.* 12 July, 2016.
https://www.washingtonpost.com/news/the-switch/wp/2016/07/12/holocaust-museum-to-visitors-please-stop-catching-pokemon-here/

[10] Rice, L. (2016). 'Pidgey's Law' would help remove problem Pokemon Go stops. *dnainfo Chicago.* 24 August, 2016.
https://www.dnainfo.com/chicago/20160824/rogers-park/pidgeys-law-would-help-remove-problem-pokemon-go-stops

It's getting more challenging to observe tech use, as much of it lives on students' phones!
Image Credit: Nick Della Mora (via Flickr-CC[11])

participate and to ensure that any technologies used to facilitate these activities are resulting in the desired outcomes.

Awareness of the balance of tech and activity. A good first question to ask for a learning activity is "what is the balance of activity?" It helps to make a mental note of all of the different things students will be doing in an activity. For example, when thinking of the Pokémon Go example, participants *walk, talk to other people, navigate, tap on their phones, chase pokémon, and challenge rivals* to battles. Most of these activities happen simultaneously, thus creating a balance of activities. Negative outcomes can emerge from this balance.

Imagining how the balance of activity could be disrupted is another useful question. Unfortunately, this requires those of us who implement technology to sometimes think like a bad guy. In today's world, both outsider and student trolls and bullies constantly try to disrupt activity just for the fun of it. In addition to thinking about how participants could throw a wrench in learning activities and drive students off course, it's useful to think about safety in all things digital. Cyberbullying is a negative aspect of today's digital ed-tech world that teachers need to consider, and the one you'll

---

[11] https://www.flickr.com/photos/nickdm/8382252948/

most likely encounter. It is easier to maintain the balance of activity if you have an idea of the activity mix going on.

Brainstorming stuff people do. If you integrate a new tech, what will students do with the tech? What kinds of activities can they do with it? How do these tech-supported activities lead to learning outcomes? All of these questions touch on a necessary point of needing to anticipate some of the common ways that students may act using tech. Some helpful questions related to finding data to help you understand what students are doing include:

- What kinds of work or creative products will be done by students (e.g., writing, images, drawing, videos, in-person discussions, etc.)?
- What kinds of analytics or participation tracking options are available in the apps I'm going to use?
- How will students work together, or work with people outside of the classroom (e.g., parents)?
- How will I know when one type of activity is different than another type? In other words, do I have a list of the types of activities that students will do?

Having a running list of what actions to look for will help you make adaptation decisions later on down the road if it isn't going according to plan.

Secondary app interactions—what are students using? The tech-based activities that educators integrate into class work may also be amplified or squelched by the other apps, software, and hardware that students use in their everyday lives. To monitor activities that your students are doing and keep tabs on any undesired outcomes, it's also valuable to consider the other apps that students are using. This is especially true if students are doing any out-of-class activities, such as field trips, homework, or blended learning apps.

The types of apps that students could commonly use that could interact with class activities include chat and messaging apps (e.g., Snapchat, Kik, WhatsApp – see Chapter 1), media sharing and re-mixing apps (e.g., Instagram, Vine), social networks (e.g., Edmodo, Facebook, Twitter), games (including Pokémon Go), and personal productivity apps (e.g., Google Docs/Drive, to-do lists). Educators can benefit if they poll their students every so often to see what apps they use and how they use them. Try to imagine how these apps may interact with what students are expected to do for class. Sometimes the negative aspects can be immediately apparent (such as with facilitation of negative feedback or an unproductive backchannel among students), but some behaviors may not emerge until after classroom activities are underway. As such, keep an eye on how students interact with classroom tech and what apps they use to help facilitate their work.

Monitoring and adapting. Educators cannot completely anticipate the different ways that classroom technology will be used, nor the many outcomes that can happen as a result of tech-based classroom activity. As such, it is important to continue to monitor and respond to changes in how students interact. Adapting is normal and expected. When technology is used in a class for the first time, it won't work exactly to expectation every time at launch (and it likely never will). So, we are left to the same procedures that the best app and hardware developers of our day use: you watch, use your observations to identify patterns, and you change to meet the needs of learners.

If the Pokémon story is any lesson, it shows that even the best apps have experiences with planning only going so far. Complexity in interactions requires educators and software developers alike to stop managing for one specific outcome and monitor changes as they occur. This is especially true with the uniqueness of needs, learning styles, and interests of individual classrooms.

The best strategy with new tech implementation is to continually assess whether the tech-based activity is aligning with the goals that were set from the outset. If you didn't have any goals in mind, take a step back and set some to prevent the tech from "driving" you and your students. It's also helpful to find ways to see what other activities students are doing in their everyday lives and how these activities might additionally benefit student achievement of learning outcomes. This may require us to be imaginative, though, as the benefits and technology interactions may not be immediately apparent. As such, don't be too hasty to shut down activity that is new or you don't quite understand. New activities and combinations of technology-based activities might lead to incredible outcomes, so keep an open mind with an eye toward goals.

17

Mastering the Search Bar

For educators, there's more than Google to navigate the sea of information

Google has staked a claim on every piece of real estate in the online world. Everywhere you turn on the web, Google encourages visitors to find what they're looking for. They promise an ease in finding information. However, despite all the advances in search technology, there's no guarantee you'll find desired information within the first two pages. This has become a challenge for both Google and for educators.

Information sources have appeared exponentially. Google, among other search providers, place hundreds of thousands of new web pages in their search index each day. Not only does Google have to make sense of all this information and decide what to share with people who use search, but searchers have to make sense of the results. Unfortunately, searchers still have to dig past the first page of results to find what they're looking for. This is especially frustrating for young searchers, who might give up after the first page. The tension between instant gratification and a spirit

How can we better navigate the stacks of resources on Google?
Photo credit: Joel Penner—Paper Weaving (via Flickr - CC[1])

of inquiry can be tough to balance when Google supposedly gives you what you want on the first page.

In this chapter, I check out some useful strategies for "getting past the first page" on Google and other providers to find what you're looking for. I also take a moment to dream of a few classroom scenarios that educators can use to encourage the use of the most appropriate search tools and to instill useful searching habits with students.

Google's main page is good, but it's not the only place to look

We all know it intimately by now. Our web travels have brought our requests to its servers thousands of times. The simple rectangle on an all-white page stares at us from our phone screen or desktop computer. If we're lucky, the logo on the page will be a clever illustration commemorating a special occasion on the day we visit.

---

[1] https://www.flickr.com/photos/featheredtar/2302651444/

The unassuming "GOOGLE SEARCH" button waits for our command, which launches Google's search algorithm to find our desires to the best of its ability. Today, it's estimated that Google has over 60%[2] of all primary internet searches performed at their site (with the runners up being Bing around 20% and Yahoo around 10%). Google search has integrated itself into almost every aspect of a typical digital life. What you may not know is that Google has a number of other services in addition to the main-page search bar that may better help you get the info you want. Also, there are a number of other services that may be helpful outside of Google that have become popular for searching out information on the web, which I examine below.

Google's other services focus on delivering specific types of information, or in certain formats. Any regular visitor of Google might have already stumbled on the common alternative searches, such as Google Image Search, or Google Video Search. These searches allow users to specify their search terms while restricting the search to a particular type of media. These might be of great value to students as they search for information related to a project, or are looking for multimedia sources.

In addition, Google also has a special search called **Google News**[3], which focuses on recent news stories from sites that have been determined as journalistic outlets. Newspapers, online news, and news video are all included in this search. Similarly, **Google Books**[4] focuses on searching the vast repository of books that Google has archived, making this search very similar to one you'd perform in a traditional library. Although the results might be a bit too dense for K-12 students, **Google Scholar**[5] is a search that can be used to generate scholarly articles and reports. Google Scholar has become increasingly popular among researchers and the general

---

[2] comScore. (2016). comScore releases February 2016 U.S. desktop search engine rankings. *comScore blog post.* 16 March, 2016.
https://www.comscore.com/Insights/Rankings/comScore-Releases-February-2016-US-Desktop-Search-Engine-Rankings

[3] App- Google News: http://news.google.com/

[4] App- Google Books: http://books.google.com/

[5] App- Google Scholar: http://scholar.google.com/

public alike due to its ease of access for retrieving scholarly works. Google has also been working to archive the world's patents, copyrights, and intellectual property claims, as **Google Patents**[6] gives searchers the capability to find detailed information and actual patent filings with a number of search parameters. Although these search engines are focused on particular types of information, both K-12 and higher education students may find some inspiration for their projects in their results. Google's specialization in these areas have really made my life easier when searching for certain types of information.

Although Google is the go-to place for search, there are a lot of other productive sources of information on the web that Google doesn't necessarily pick up or prioritize. Google focuses on web pages and not smaller nuggets of info like tweets or Facebook posts, so some other sites may be helpful in giving you a second opinion or a fresh source of information. The **Twitter Advanced Search**[7] function gives investigators a great tool for finding up-to-the-minute comments and web links that have been shared by Twitter users. I find myself using this frequently if I am looking for timely information, or even articles that are not yet appearing in Google search. If you are following a particular hashtag, you can search it in the advanced search page on Twitter, or you can use a number of services to help you keep track of live discussion around the hashtag. Having a "live feed" of a single hashtag is useful when you are following a Twitter Chat or live event that uses a hashtag or specific word. I've used **TweetChat**[8] for real-time hashtag search, and have found it to be a good service.

Other places where people gather information are also valuable places to search. Next time you're looking for something, try using the search at the web archive **Delicious**[9] to look through thousands of users' public web bookmarks. You can also try a search at

---

[6] App- Google Patents: https://patents.google.com/
[7] App- Twitter Advanced Search: https://twitter.com/search-advanced
[8] App- Tweetchat: http://tweetchat.com/
[9] App- Delicious: https://del.icio.us/

Reddit[10], one of the largest communities of specialized interests on the web. Reddit hosts thousands of specialized pages called subreddits that span all topics and interests. A search on Reddit is bound to turn up some useful results. The question-and-answer site **Quora**[11] can be useful for people searching for answers to questions others have already posed. Finally, if you are looking for information on anything technical, the large community at **Stack Exchange**[12] has an archive of a countless number of discussions on anything ranging from computer issues to programming to statistics. Although I've listed a few here, there are many other specialized communities and message boards on the web that you can individually search for more directed results.

Google Alerts[13] are another search-related option for long-term projects. Savvy users can set up an alert to be sent to email (or to an RSS feed) anytime new sources with the search terms are found by Google. Check this out—it can be a real timesaver! And, it can give you stuff each day that would never make it to the first page of google.

Thinking pedagogically about web searches and student inquiry

Getting students to regularly interact with web resources is one of the best ways to instill productive search habits and information literacies. There are a number of fun ways that teachers can coach student inquiry in any number of projects that they work on.

Lesson plans from Google Educators. As luck would have it, an interested educator who wants to bring more thoughtful internet

---

[10] App- Reddit Search: https://www.reddit.com/search
[11] App- Quora: http://quora.com/
[12] Service- Stack Exchange: http://stackexchange.com/
[13] App- Google Alerts: https://www.google.com/alerts

148   Mastering the Search Bar

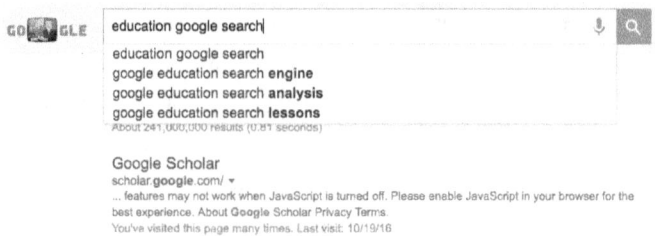

*There are many ways that educators can use Google
in refined ways to help their students succeed*

searching and inquiry into their classroom can enjoy a variety of lesson plans[14] that have been developed by Google Certified Educators and the Search Education staff at Google. The search team at Google also released a webinar video[15] on how to integrate search into classroom activity. These lesson plans have useful activities for the proper execution of search commands and technique. Because web searching is universally valuable for any project, these lesson plans can be useful for any subject area. Activities in the lesson plans can be adapted to match the types of things students are studying or the activities in which they are engaged.

Describing what we want. Almost every search algorithm is built on the idea that we will use combinations of words to describe what we want to find. As we know, similar concepts are often described by people in different ways with different words. Sometimes, we may not know these alternative words or phrases, or have thought of every possible way to describe our intended search. An activity with students that deconstructs the concepts or events that are being investigated to find keywords, synonyms, and alternative ways to describe the item of interest can be helpful for finding information on search. Teachers may give some "starter"

---

[14] Service- Google Certified Educators Lesson Plans for how to search
https://sites.google.com/site/gwebsearcheducation/lessonplans

[15] Google Videos (2009). Teaching search in the classroom webinar. YouTube. http://www.youtube.com/watch?v=X4FXXMzUiyo

keywords to help students on their information-seeking trek, akin to providing a vocabulary list in a class. In fact, vocabulary may be one of the most important skills in web search! Instead of giving away answers, teachers may also choose to use classic tools like a thesaurus or dictionary to develop alternative descriptions, which many are readily available on the web.

Wikipedia miners. Wikipedia does have its uses! Although some people question the reliability of crowdsourced resources like Wikipedia, the articles on the site are excellent sources for finding keywords. In the sense that people who browse wikipedia are "digging" through it to find useful information, educators can work with students to mine Wikipedia articles for helpful nuggets (think pickaxe and minecart). Because the site is designed to be a helpful first-stop for people doing research, the articles are full of vocabulary and ways of describing concepts under investigation. In particular, attend to the special nouns, adjectives, and verbs. Look for stuff that stands out and specifically describes the subject of study. Activities that dissect a Wikipedia article can produce a list of useful keywords for further search!

Coaching for habits of search. It's ok to help students with some starter keywords or give some specific instructions on what to search. Even though it's more work, it could be also helpful to have students puzzle out their list of keywords and where to search. The role of the teacher in internet research might be best suited then as a coach instead of an instructor. Teachers give students advice on things they find, but the students continue to drive the research. Open-ended research questions and search possibilities give students practice with the often ill-defined questions that people have to make everyday, as well as evaluating the resources from searches to answer those questions. Don't just tell students what to do and have them memorize procedures. Instead, coach them and coax them into finding resources on their own from the very first step of the research process.

Peer evaluation of resources. Get everyone involved in judging the quality of work by having peers evaluate sources together. In groups, students can each take some keywords from a list generated by the group and return to evaluate the results. Students can also pair up to evaluate others' findings, as well as determine the strengths and shortcomings of each source. Although this activity might only marginally increase the efficiency of searching, it would also give some bonus practice time for working on evaluation of source quality and experience with finding resources in various places.

Many-tab mayhem. Instead of clicking on each result in a search individually, open up links in new tabs. Stay on the search results page for a while to reduce being distracted by resources you find. By staying focused on your initial search parameters, students can better remember what their goals for the search was. I find myself personally susceptible to losing track of where I was in a search if I follow results immediately. Instead, I open about 10–15 tabs of things that look interesting based on the search result headline alone. Only after I have opened many tabs on a subject will I now start to evaluate what I found. As I find something that doesn't fit my interests, I just close the tab. When I am done with the mayhem of tabs that I opened, I continue with a new search or pick up the old one where I left off. I find myself being influenced by information review as well, picking up new keywords and combos to try as I read through the media I find on my collection of tabs. Many browser tabs might be clutter, but in my case it can be an organization tool!

18

Toward Criteria for Evaluating the Quality of Digital Information

Useful habits for evaluating web-based information

I've had a lot of conversations lately about what I consider to be "quality" information from the Internet and how to go about finding the elusive "quality" article or web link. Because everyone can publish anything they want, does that immediately make everything unreliable? Does something have to go through editing or peer review for it to be quality? There's gotta be some ways to make our lives easier when searching out the gold nuggets of info on the web.

Is the age of the "unbiased" source gone, and are inherent biases something that we just need to deal with today? If we want students to become savvy users of internet information, we need to instill a culture of critical evaluation of things found on the web. But the criteria for evaluating information can be elusive In this chapter, I take a few moments to tackle some of these questions and provide some thoughts on the criteria I use for when I am searching out in

Where to we even begin to start with evaluating the quality of info on the Internet? Photo credit: Web 2.0 Icons via Flickr[1]

formation on the web. Although I quickly brainstorm some of the criteria of quality that can be used, each of these criteria could have a chapter of their own. As such, it's worth exploring a bit more on your own if you're interested in a particular criterion I have chosen.

What does "quality info" even mean today?

To me, the word quality involves some sort of judgement call: you're comparing the object in front of you to some other object or group of objects you've experienced in the past. When thinking about a quality shirt, a quality TV show, or a quality YouTube clip, we think in terms of good and bad, not necessarily if it's fact or opinion (although that may go into our judgments about quality). So, this term immediately makes me think that this is a judgement that will be different from person to person.

To start to think about this term, I looked at a few dictionaries (I found this a useful exercise while thinking about my own assumptions about the word "quality"). In an evaluative sense, Merriam-Webster's defines quality[2] in the way I assumed above: "how good or bad something is." However, their second definition adds a

---

[1] http://www.flickr.com/photos/7363465@N08/3925513417
[2] Merriam-Webster Dictionary (n.d.). Definition of 'Quality'. http://www.merriam-webster.com/dictionary/quality

bit of nuance to the term, thus making it a bit more complex: "a high level of value or excellence." High level of value...what does that mean? Oxford's definition[3] gives even more leeway to the person doing the evaluating, saying that quality is "the standard of something as measured against other things of a similar kind or a degree of excellence in something." So when we use the word *quality* to describe digital resources, we're likely talking about something that is good and has high value. However, that value and "goodness" is going to be different for everyone. The "one person's junk is another's treasure" saying rings true here.

So, we not only have to think about what "good and valuable" mean, but we also have to think about how others interpret the same terms. Many an internet argument has started over what one person passes as acceptable, quality information with another passing the same as junk. This is exactly why we need good ways of thinking about quality: so we can evaluate and defend our choices of information that we use to learn and make sense of the world.

Questions about quality have been around for as long as we've been using language and are certainly not new to the digital age. However, with new media channels and new types of storytelling that are emerging every year, it's useful to reflect every once in a while to evaluate how do we go about assessing quality in the types of media and information we use. These criteria will likely change with the times as the use of communications media gets more complex and the tasks for which we use information shift.

Before I get into a discussion on quality criteria, though, it's a good idea to ground these ideas with an assumption that anything that was written anywhere was written by someone. This extends past the traditional "writing" of texts and includes authors of videos, audio, images, and other media. Thus, with each of us as humans who experience the world differently than one another, we have inherent biases in our communications based on how we see and describe the world. These often come across as differences in what information we include, what information we exclude, and how we order information in the stories we tell. Even the most unbiased news sources have some sort of approach to their storytelling

---

[3] Oxford Dictionary. (n.d.). Definition of 'Quality'.
https://en.oxforddictionaries.com/definition/us/quality

that could be considered a bias. This should be accounted for when we think about quality in information

Toward criteria for digital quality when teaching digital literacy

It is increasingly important to promote the development of digital and information literacies in students. The ability to translate meaning across various media for different projects is an important 21st century skill that will increasingly become necessary as media become more complex. Along with the use and interpretation of digital media comes the need for the ability to discern the usefulness of information, in whatever form it takes on the internet.

Interestingly, most of the resources I found when compiling this list were from university libraries aimed toward undergraduate students. I didn't find very many K-12 resources out there, with only a few blog posts coming up in the first many pages of Google. I'll include a list of these sources in a section below.

Leave it to researching a topic about finding information to end up finding a ton of information on finding information! (...that sentence made sense in my head)

I'll not get too much into the details here, as I don't want to simply repeat others' work and thoughts on the matter and pass them as my own. However, I've grabbed some of the big strategies that are generally being used to evaluate internet information and ordered them a bit here. If I contribute anything extra to this topic in this chapter, it's that I argue for organizing these criteria in a way that saves us, as internet info evaluators, time and cognitive effort when performing internet work. Many articles I found prescribe a 10-step critical process for evaluating every internet resource we find, but realistically nobody has that kind of time. Instead, it may be useful to quickly evaluate on a couple general criteria, and get into the nuances when needed.

There are two main categories that I think most of the criteria for evaluating digital information (or really, any information...there's just so much that's digital now). The first is *relevance*, which relates to the assessment of whether something is valuable or good for the purpose. The second is *validity*, which is used to evaluate whether the information is correct or reliable. Now that I've got these ideas on paper, I find myself gravitating toward thinking about these two categories of most of the time when I make snap judgements on information. It takes practice, but eventually you start to ask quick questions in your head and spot issues related to these two criteria. In other words, I try to rapidly ask myself if 1) a resource is valuable to me for what I'm doing (relevance), and 2) whether it is *right* (valid and correct). Although there are many strategies used by people when evaluating info, I saw that all of the criteria that are used generally fit within these couple of categories. Asking these two questions capture the gist of much of the other criteria commonly used to assess info.

So, without further adieu, here's a roundup of the criteria I found for evaluating information, and my suggestion for lumping them into two general questions about quality.

Relevance: criteria on how valuable something is to the researcher for their goals.

- **Value**. Thinking about quality depends on your goal. Ask yourself "what are you trying to do?" and think about what kinds of information would be useful for that goal. Information is inert until it is used for something. Gaining knowledge of facts can be fun in its own right, but is truly valuable when you're using information for your pursuits.

- **Speed and timeliness**. Is the resource you found outdated? Is it relevant to current events and trends? Does the resource have the most current information? Older resources can be ok too, as long as they're relevant. If it's older, does it provide a helpful unique perspective that may not longer be mainstream?

- **Authority**. Does the resource rely on authority to promote its quality? Does it scream "listen to me! I'm an expert!" Is the information resource being written by someone who considers themselves an expert? If so, what does it mean to be an expert and how much weight does that have for your goals? Expertise is good, but if a resource rests on just claims of authority or expertise on the matter without any additional substance, it may not be helpful to your ends.

- **Aesthetic**. This may seem a bit shallow, but does it look nice? The aesthetic of an information source can matter in how people interpret information and its relevance. Does the resource have the form and shape that is expected? Is it *pleasant* to view? If not, it can be a deterrent to using the resource. This is particularly important with visual and aural media. Multimedia resources are more successful at being relevant if they're assembled in a pleasant way.

- **Meaning**. In multimodal communications (e.g., combinations of text, video, audio, games, interactive animations and graphics, and other images), there may be many embedded meanings for different audiences and purposes. The resource may not have been developed for the purpose for which you'll use it. Ask yourself how the medium is used communicate for this purpose and how ideas are depicted to get an idea of the meanings of the media as intended by the author.

<u>**Validity**</u>: criteria on how correct or reliable something is based on its claims.

- **Accuracy and factuality**. This is harder than it seems when you dig deeper into what a fact actually is. It goes deeper than asking whether or not the information depicts things *actually happened*. Instead, there are often multiple contrasting factual accounts of something, as different authors have differing definitions for describing the presence of phenomena, concepts, and events. Facts are also often balanced with opinions. In some instances on the web (particularly in the case of a Wikipedia article's talk page[4]), the community that develops an information source debates what constitutes a fact or how to define and describe events. Sometimes,

---

[4] Wikipedia. (n.d.). Using Talk Pages. Wiki article. https://en.wikipedia.org/wiki/Help:Using_talk_pages

complex events take multiple people to describe, such as in the case of an encyclopedia chapter or news article.

- **Argumentation structure.** Even in an information source that lists "just facts," the authors are always making an argument of some kind. It can be as simple as "these are facts, and they matter." The strength of arguments matter to the validity of an information source. This, however, this assumes that everyone's following the same rules of argument logic. It seems on the Internet that people often don't get too obsessed with the formalities of argument structure during a Facebook politics debate on a friend's page! However, there are a lot of resources out there[5] on the reason for argumentation and how to teach it, so I won't get into it here :).

- **Omission.** What has been left out of the information resource? This can be tough, because you have to know enough about something to know something isn't included. However, the omission of information in a resource is an indicator of the inherent assumptions and biases of the author and can be related to the resource's validity. If anything, it's useful to remember that all information sources are incomplete and don't contain every voice, story, or fact. This can help you keep a critical eye on the sources you use.

---

[5] There are numerous starting points for finding resources on evaluating argumentation structure and quality. I'll list a few here.
(1) Google scholar search of "argumentative writing,"
https://scholar.google.com/scholar?hl=en&as_sdt=0,14&q=argumentative+writing
(2) NSTA rubric for evaluating scientific explanations,
https://www.nsta.org/conferences/docs/2015SummerInstituteSecondary/ScientificExplanationExampleRubricFinal.pdf
(3) Edutopia article on evaluating scientific argumentation
http://www.edutopia.org/blog/science-inquiry-claim-evidence-reasoning-eric-brunsell
(4) Presentation slides by Joseph Krajcik on supporting students in their construction of scientific explanations
http://umaine.edu/center/files/2011/06/Bagor-presentation-Explanation-V4.pdf

- **Ordering and sequence.** Just as much as the inclusion and omission of information is important, the ordering of information in the resource can matter on how you use it. Linear media like videos, audio, and texts can influence readers by loading information up front or toward the back. The ordering of information matters and is a useful thing to think about when it comes to validity.

- **Origin.** Where does the information come from? This is also called "provenance" in some academic lingo in the information sciences. Does it matter where the information came from and who wrote it? Is it anonymous, and if so, does that matter? Knowing who published a resource can help indicate to the critical information consumer some key things about their goals and assumptions about the information they are communicating. This isn't to say that everyone has ulterior, devious motives to deceive people. It's only a useful habit to think about where authors are coming from in providing their information and how it relates to the validity, correctness, and completeness of the information you found.

- **Feedback mechanisms.** Although this may not be an indicator of accuracy or quality in itself, the presence of some kind of feedback process or editorial review is evidence that an effort was made to improve the quality of the resource. Another set of eyes might help improve the quality of documents, as a single person might not see all of the areas for improvement. Review from others may not improve the quality at all, and may even slow down publishing or change the original intent of the document. But, the presence of collaboration could be an indicator of quality for a resource as it went through an additional intentional process to improve the quality. However, don't discount something that hasn't had an editorial process, though. Some of the best resources, stories, and media are those that were first posted by the author and then subsequently received feedback only after publishing. As such, another indicator of quality is whether there are formal methods for providing feedback to the author, such as comments, a form to submit feedback, or even the author providing their contact information. Excellent conversation about a resource sometimes unfolds in the comments or talk section on a resource.

Additional Resources

Some university libraries' lists of criteria for evaluating digital resources

- Arizona
  http://www.library.arizona.edu/help/tutorials/webinfo/

- Georgetown
  http://www.library.georgetown.edu/tutorials/research-guides/evaluating-internet-content

- Illinois (Urbana/Champaign)
  http://www.library.illinois.edu/ugl/howdoi/webeval.html

- Johns Hopkins
  http://guides.library.jhu.edu/evaluatinginformation

Some K-12 blogs and articles on how to evaluate digital resources

- Strategies and activities for teaching information evaluation (Educause)
  http://www.edutopia.org/blog/evaluating-quality-of-online-info-julie-coiro

- Lessons from Scholastic.com on information literacy, evaluation, and plagiarism prevention
  http://www.scholastic.com/teachers/top-teaching/2010/11/reliable-sources-and-citations

- TeacherTap criteria for evaluation
  http://eduscapes.com/tap/topic32.htm

19

Combating Cyberbullying in Classrooms

Keeping an eye out for bullies and trolls when digital tools mask their behavior

In the age of social media, YouTube accounts, and messaging apps, people have been able to connect in far more ways than years past. However, new forms of communication have brought new tools for bullies to hurt others. As teachers in modern, digitally driven learning contexts, we must keep our eye out for new types of antisocial behavior that can be hurtful and damage students' abilities to learn.

Known also as **trolling** and **griefing** in various online circles, **cyberbullying** on the web has a powerful influence over students. The spread of gossip and negativity can really hurt a student both in an out of classrooms. In fact, cyberbullying isn't limited to just students—teachers can face it as well! As internet technologies have proliferated, bullies have found new domains to pester and hurt others. With anonymity on the web and a greater opportunity to expand learning to outside of the classroom, both internal and external bullying can negatively influence learning opportunities, as well as hurt students in more ways than just their scholastic achievement.

New technologies, antisocial behavior

As new communications technologies have taken over education and students' personal lives over the last 10 years, the landscape of bullying has also changed. Although bullying has been around for as long as anyone can remember, the web age has amplified its ways of reaching students and potential for harm.

Classic forms of cyberbullying, such as negative comments to others, are amplified in online spaces. They tend to persist past the classroom because they can "jump" to other apps that students use as students communicate with each other outside of the traditional classroom walls. In other words, bullying can happen in classroom-sponsored apps, as well as those that students use every day. With social media, damaging gossip and rumor creation can spread like wildfire and cause miserable experiences for victims. We gain many new opportunities with digital technologies, but we as educators must also be aware of the types of negative activities that can occur with these tools.

Keeping an eye out for bullying with digital learning tools

There are many things to keep an eye out for when you use certain pedagogical elements within your classroom activities. Bullying behavior can hide from teachers' sights and is often obfuscated by bullies to prevent any recourse. However, there are some things to watch for in common activities using digital resources.

Feedback and collaborative work. Watch for negative, not-helpful comments from students when they participate in a feedback-giving activity or collaborative project. It's important to make sure you spot unproductive comments as they happen. This would require teachers to see the feedback that students provide. Comments that are not private to the teacher would work best, but that doesn't stop students from talking about others' work in non-school apps that they use.

Social networks. Facebook, Twitter, Edmodo, and many other networks are popping up in classrooms. Because it is so easy to share information, it can also be easy to spread gossip and rumors. Keep an eye on the channels being used by your students for class work. Also, it's not just text: hurtful photos, videos, and audio can all be easily shared on social networks. Students are also likely on networks that you can't monitor, so this may spill over to classroom projects as well. Students also can use social networks and messaging apps to open "backchannels" during class, allowing them to have conversations during formal class time. This is hard to monitor, but teachers should try to be aware if conversations and sharing of information is happening, especially with apps that are formally brought into class projects.

Private messages. Many apps provide the opportunity to send private (PMs) or direct messages (DMs). These could provide opportunities for bullying behaviors because teachers can't see. It's important to know if apps you use have any features where students can send private messages. Although teachers won't likely be able to see messages, it will give a clue to watch for any discussion among students about negative things being shared privately.

Anonymity. A challenge of the web that has empowered cyberbullying is the anonymity that some apps provide. Try to find apps that provide some kind of identification so you can know who is posting what. If participation from the public is allowed, try to limit it or diminish the negativity that anonymized posts can provide. This may require some moderating time on a teachers' part to make sure that discussion and sharing remain positive and constructive.

Things to do if you spot bullying in digital spaces

Although you can't always prevent bullying behavior from happening in online spaces, you can certainly try to identify potential areas in which bullies can affect learners. Some preventative work and monitoring can go a long way in providing a safe learning environment for students as they use digital technologies.

Know what you can control. When you use a new web-based technology in your classroom, you should learn what is public versus private, including if your students can send private messages to each other in the tools you use. Also, just because you have control over certain limitations in the "official" tools you use, you may not have complete control over how your students interact. Your students will likely use secondary apps of their own to communicate, so it is wise to have an idea of what kinds of apps your students use.

Set rules for moderation. You should also find out what kinds of controls you have on individual and group accounts. Although the most open policies can do wonders for organic conversation and group dynamics, some controls may become necessary if you witness students being mean to each other. These types of controlling actions are typical of moderators in an online forum. In addition, many websites that offer social or other interactive functions have established rules of conduct that determine what kinds of behavior are allowed. It may be wise to set rules of your own with your students ahead of time for things such as language, mean behavior to others, and where the line between productive disagreement and bullying is drawn.

Control what you can. As a teacher, you have some control over the activities that your students do. If you spot internal bullying within your class in online spaces, do what you would do if it were face-to-face: take it seriously and put an end to opportunities for bullying. With digital tools, this may involve locking down some of the functions on the tools you use or altering the privacy settings for some users. For an internal bully, schools may also have certain disciplinary procedures that should be followed. Consult your school for what should be done when you observe bullying behavior in your class.

Don't feed the trolls. Digital *trolls* seem to only take joy in others' misery. They often hide behind the web's ability to stay anonymous and pester legitimate users of apps, forums, and other social spaces. One of the primary rules of internet interaction is "don't feed the trolls," or don't do anything to fuel the interaction with a negative participant. This is especially true if the participant is anonymized or not a part of your classroom. If you can, try to block or close off your activities from potential trolls. However, some activities are more beneficial if activity is public (such as crowdsourcing or sharing one's work), so it is good to know what could be possible if you open work up for public interaction. Learning how to properly deal with the trolls on the web will likely also become a critical digital skill for success in the web-based world.

Additional resources for learning more about confronting cyber-bullying

- **Stopbullying.gov: What educators can do.** A useful website with many resources on how to spot and react to bullying behaviors in classrooms, as well as out-of-school spaces.

    https://www.stopbullying.gov/what-you-can-do/educators/

- **Common Sense Media: Cyberbullying toolkit.** This is a useful set of resources for helping you identify cyberbullying when it occurs and how to respond in a productive way.

    https://www.commonsensemedia.org/educators/cyberbullying-toolkit

- **MediaSmarts: Resources for teachers.** Media literacy plays an important role in combating cyberbullying. This site provides resources on the different kinds of bullying behaviors that occur with various media, as well as some tips for teachers on how to use different digital tools, apps, and media most effectively.

    http://mediasmarts.ca/cyberbullying/resources-teachers

20

BYOD: Bringing Students' Mobile Devices into the Classroom

Using the devices students use every day to connect and engage

According to a 2015 Pew Research Center study[1], more than 75 percent of teens have personal regular access to a mobile device that is connected to the Internet. In the same study, 68 percent of teens use the Internet on a daily basis. However, this is not necessarily "logging on" in the conventional sense—it is mostly via mobile devices and apps. In fact, students in grades 6–12 personally use phones and tablets more than computers, laptops or the internet at home[2], and more young people tend to own a phone than older

---

[1] Lenhart, A. (2015). Teens, social media, & technology overview 2015. *Pew Research Center: Internet, Science & Tech.*
http://www.pewinternet.org/2015/04/09/teens-social-media-technology-2015/

[2] Smith, A. (2015). U.S. smartphone use in 2015. *Pew Research Center: Internet, Science & Tech.* http://www.pewinternet.org/2015/04/01/us-smartphone-use-in-2015/

folks. Some tech observers have even gone so far as to argue that your mobile phone may be the only computer you will own in the very near future[3]. Although I think we're still pretty far away from abandoning our desktops, tablets, and laptops, it's safe to say that mobile devices are here to stay and have become important parts of life--including for K-12 and college students who have grown up with these devices. As the price of data plans and phones has dropped over the last five years, cellular data networks have brought the internet to many households in which there is no hard-wired broadband access.

Because of this, it appears that the next digital divide will not be caused by a lack of access to technology, but instead of a lack of skill for using technology in principled ways. By not integrating the tools that students use in their everyday lives, educators risk a lack of alignment of daily activities to the tools and skills that are necessary to complete them. School activities prepare students for the rest of their lives. Technology should be integrated in principled ways using the tools that will be necessary for future success.

It's time to think pedagogically about the devices that students use every day to access the internet and live their lives. Mobile devices have become the tools with which students use to process the world. Because devices have become so ingrained in students' everyday lives, it is essential we use the same tools they are familiar with. If not, educators risk taking away the very tools that students use every day to process the world.

Instead of requiring students to use classroom- or school-owned technologies, such as a computer lab, teachers are making innovative use of students' own pocket devices. Many schools ban the use of devices during the school day as they can be distracting to learning activities. Although this is true, a growing trend at schools called **Bring Your Own Device (BYOD)** promises to make principled use of students' personal technology in ways that they understand and are relevant to their daily lives. Although access to technology remains an issue (e.g., some students don't have phones), a dramatic reduction in access can be imagined if educa-

---

[3] Bonnington, C. (2015). In less than two years, a smartphone could be your only computer. *Wired Magazine.* 10 February, 2015.
http://www.wired.com/2015/02/smartphone-only-computer/

tors shift their paradigm from the computer-on-a-desk ideas to computers-in-the-pocket.

Thinking pedagogically about students' own mobile devices

There are many things that you can do with the tools on mobile devices. Challenge yourself to think about each of these functions pedagogically, or Some of the questions on thinking pedagogically with mobile devices include how might you involve them in a learning activity, what new things are students doing with the tools compared to other methods, or how are the experiences created memorable and give students a "finished product" that they can keep? Below, I discuss some of the primary activities that can be done with most mobile devices.

Communications. This is the first inclination most people have when they think about what a mobile device can do. It's a phone, messaging device, and social media hub. Apps like **Edmodo**[4] are social networks developed for education and are generally closed to the public. Facebook also offers private groups, as well as google communities. K-12 and college students alike are also likely using messaging apps like **WhatsApp**[5], **Kik**[6], and **Snapchat**[7]. Also, don't forget about all those social networks that students may be on like Facebook, Twitter, and Instagram.

However, it's important that you be mindful that your students may not want to mix their personal and school lives, or put their work out for the world to see on social media. Keeping students' personal lives in mind, think of ways that you can creatively use social media and communications apps by creating group or class projects, closed networks, or class messaging sessions. Some students may find it fun that you use the apps they are familiar with, some may not. It's worth having a conversation about these boundaries and what would be cool to do with the class with their favorite communications apps.

---

[4] App- Edmodo: https://www.edmodo.com/
[5] App- Whatsapp: https://www.whatsapp.com/
[6] App- Kik: http://www.kik.com/
[7] App- Snapchat: http://www.snapchat.com/

One useful activity for communications apps and instant messaging is to create a class "backchannel" where students can post burning questions or comments as they have them. Although this

This is a QR code. Try downloading a QR reader app and scanning this code! Find a QR reader app by searching for "QR reader" in your app store.

can seem distracting, it can be used in a counterintuitive way to help students focus and generate useful lines of inquiry.

Digital storytelling using a multimedia production studio. Today's modern mobile phones, both iPhone and Android, contain enough tools to be multimedia production suites. Phones typically have cameras that can record video and photos, audio recording capabilities, and the ability to edit and share these creations. Apps like Snapchat and YouTube are gaining popularity among teens and college students for storytelling among friends and family.

Again, it's important to note that educators should be careful not to hijack these apps for educational use and expect the same level of interest from students. Any use of an app for educational purposes will not always align with how students use apps for social purposes. As such, it is not prudent to try to mesh these two worlds together. Instead, we can think of creative ways on how students can make the most of these apps for storytelling and to share their work with other students and potentially a broader audience over the internet.

Interactivity with other objects. Using digital scanning apps, such as QR code scanners, educators can make the physical space of the classroom, the school, or other public areas come alive. QR codes (short for Quick-Response, see example below) allow students to scan a barcode-type image, which will take them to a link on the internet.

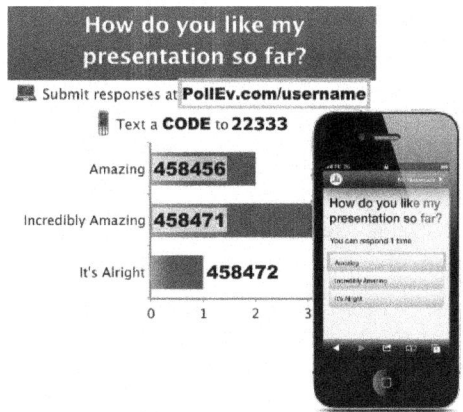

A sample of a mobile phone/internet poll on a projector using the Poll Everywhere app. Photo credit: polleverywhere.com

Teachers can create QR codes and print them out using apps like this **QR code generator**[8], which when scanned can lead students to links on the Internet. You can download free apps that will scan the code and take you to the link embedded in the QR code. Posters, gallery walks, dioramas, and other displays can become interactive with a QR code. Similarly, augmented reality apps (see Chapter 9) can provide additional information "superimposed" on physical spaces, which can enrich students' experiences.

Polling and student feedback. Students can share feedback with each other or with teachers using their devices in the classroom. Teachers can use apps like **Socrative**[9] or even **Google Docs**[10] for soliciting feedback from students. Using polling or "clicker" apps like **Poll Everywhere**[11], teachers can create polls with multiple choice questions to get immediate feedback from students during activities that can be responded to via the Internet or over text message.

You can also put a live poll up on a projector for everyone to see. It is a great way to do formative assessment without kids feeling too much pressure as results are anonymized and aggregated.

---

[8] App- QR Stuff: http://www.qrstuff.com/
[9] App- Socrative: http://www.socrative.com/
[10] App- Google Docs: http://docs.google.com/
[11] App- Poll Everywhere: https://www.polleverywhere.com/

Keeping students involved without mobile devices. It's likely that every student will not have a mobile device. There are ways to work around this as well. You can find some devices to borrow from other teachers, from other departments, or from IT help. You can also use other devices in a classroom, such as iPads, laptops, Chromebooks, or desktop computers to get non-BYOD students involved. By breaking students into groups, some activities may also not require everyone to have a digital device. Group roles such as notetaking, leadership, artwork, design and organization all can be done using non-digital methods. If there are classroom devices, it may also be helpful to find creative ways to find times to "switch hands", or pass roles or devices on to other students to keep everyone involved.

21

Finding Free, Web-Based Teaching Resources

Gathering open educational resources when resources are thin

I get it, nobody wants to pay for the apps we use every day. However, there is always a tradeoff when using free technologies. Free doesn't always mean the same to everyone, especially with digital technology. This isn't necessarily a bad thing, but knowing about the meanings of free can help you protect your students and practice proper web etiquette when using shared resources.

When you are looking for digital resources, such as images, videos, curriculum, lesson plans, or articles and books, it is always a good practice to try to find the license under which the resources are available. There are generally two types of digital resources that are often regarded as free in terms of financial cost: creative works/media and software. I'll discuss what "free" means in each of these types of resources below, and what educators should consider when working with them.

Type 1: Creative works, media, and the creative commons

You'll find that many people (especially educators) are very generous on the web, offering their resources for free of charge and encouraging them to be shared. However, most of the time, free does not mean you own the resource. In most cases, you will always need to reference where you found a resource. This is especially true if you are making your own curriculum and distributing it on the web. The exception to this rule is in the case of public domain licenses, which are open to everyone to use and are "owned" by the public.

You may have heard that you can use anything on the web, as long as it's for an academic use. However, there's a fine line between "academic use" (also known as **fair use**) and copyright infringement. Instead of having copyright issues arise, it's always best to stay on the safe side and provide attribution whenever you use someone else's work. It's the same idea as when writing term papers or academic articles; citing others' work is good practice and just the right thing to do. Also, students and those who use your work additionally benefit from knowing your source materials.

It's also likely that you'll encounter works that are "open licensed," typically with a **Creative Commons license**[1]. All this license says is that legally you are required to cite or "attribute" the author in any project that you use the resource. There are sometimes other terms also attached to Creative Commons licenses, such as non-commercial (NC) or share-alike (SA). CC licenses are often shortened to acronyms like CC-BY, or CC-BY-NC. Look for the license on resources you find to know exactly how you can use it.

It's also wise to license your own work when you share it on the web. If you want to retain all rights on your work, a simple copyright notice would suffice. However, if you would like to share your work and encourage others to use it openly, consider attaching a Creative Commons attribution (CC-BY) license to your projects. This legally requires others to attribute you as the author for your work while keeping the work free to share and use.

---

[1] Service- Creative Commons: http://creativecommons.org/

A sample Creative Commons CC-BY license icon. When you see this image, you can use the content free of charge without written permission, as long as you attribute the creator of the content.

Making a simple declaration like this[2] will encourage others to use and share your work around the world. Many digital resources can be released with a Creative Commons license, including text and articles, images and photos (I personally license my photos CC-BY on Flickr), videos, audio, and slideshows.

Type 2: Free software

Despite being "free" of cost, there are still strings attached to many digital resources. However, openly licensed resources are the most flexible in allowing you to do what you want in your classroom and adapt resources to meet your needs. Free software, on the other hand, often comes with additional constraints, especially if the company has not licensed the software as "open source." In these cases, the company maintains and runs free-of-cost software, often with restrictions on how you can use it and with terms on how user data are collected and used.

There are two types of free-of-cost software: those that are open source and those that are closed-source. Most educators will encounter the second kind of software, closed-source.

I'll start with the first type, which is the most flexible (and rarer). Some software is provided with an open license (typically GNU or Apache), similar to the creative commons licenses. Software with open licenses typically require you to attribute the software whenever you use it in work. Some other restrictions apply, but they typ-

---

[2] Creative Commons has a good article on how to properly attribute your works with a CC license, as well as how to properly use content that has been licensed as such. https://wiki.creativecommons.org/wiki/Marking_your_work_with_a_CC_license

ically don't affect educators who are just seeking cool technologies to bring new activities to their classes.

Some nonprofits and granting agencies (such as educational research agencies like IES[3] and NSF[4]) often develop or fund software for various purposes that are open licensed. In addition, some software projects such as **Linux**[5], **WordPress**[6] (a blogging and website platform), **MediaWiki**[7] (the software that powers Wikipedia), and **Moodle**[8] (a learning management system for online classes) are all open sourced and have large communities that work together to improve the software. The principle of open-licensed software is similar to open-licensed resources mentioned in the section above: people have some cool things they've made and they want to share. The only real difference is that software open licenses cover additional legal issues that are unique to software.

However, the dominant form of free software in education is the type that is closed-licensed. These are any software that does not cost anything and is maintained by a company. Many web-based networks and apps fall in this category, including Facebook, Edmodo, Twitter, Google, YouTube, and even most free online games……pretty much everything we use every day. Companies may offer apps for free, but they always typically come with a "terms of service" that dictate how you can use the software. In addition, companies also typically collect data on how people use their software, as well as the contributions made in the software. Status updates, posts, documents, and other things created in apps are frequently owned by the software company, or at least accessible by the software provider. One of the biggest tradeoffs of "free" software is the ownership of data and what companies can do with the data that are collected in software. It is becoming increasingly important to consider how data are used by software providers, especially in the case of students' needs for secure and private spaces for learning.

Data security is also an issue to consider when thinking about free services. Some software providers go to extra lengths to keep

---

[3] Agency- Institute for Educational Sciences: http://ies.ed.gov/
[4] Agency- National Science Foundation: http://www.nsf.gov/
[5] Agency- Linux Foundation: http://www.linux.com/
[6] App- WordPress: http://www.wordpress.org/
[7] App- MediaWiki: http://www.mediawiki.org/
[8] App- Moodle: http://www.moodle.org/

the data of their customers secure. However, it is important to remember to balance the use of free software with the security needs of your students. Always use secure passwords and avoid putting students' discussions, data, and personally identifiable information out on the open web. If the software or app that you're considering has a history of not securing their users' data, it may not be the right one for your class. It pays to do some research on apps, even though the price tag is nice.

Sometimes, app and software providers will provide better options for a small subscription fee, including no ads, tech support, more space and features, and the ability to export your data. These are features that might be worth paying for over the "free" version of an app.

Thinking pedagogically about "free" stuff

There's always a tradeoff when working with free software and resources.

Always consider who will be collecting, owning, and possibly using your students' data. Companies often sell data and the things they find out from data for advertising purposes. If you adapt software or resources to meet your classroom needs, always make sure you attribute the authors in the ways determined by the license.

Weighing all of these considerations, you may find there's more to the cost of digital resources than you anticipated. If that's the case, don't throw away resources right away. Instead, just take a moment to think a bit about what the software or resource brings to learning. If you can replace the resource with another option, that may be the best bet. However, if you get some unique features from the resource, it is likely worth it. The non-financial costs can surprise people, so just be aware of what you're signing up for.

If you can't find a license for a resource, you can usually assume it is "all rights reserved." If you are unsure, you can email authors or customer support to find out. They may be even willing to extend a license to you as an educator if it is going to be used in a classroom.

Index

Actions technology allows, 2
Adapting technology to meet needs, 142
Add-ons, 37
Adventure (Game), 16-17
Affordance, 2, 29, 46, 72, 76-77
AI. See Artificial intelligence
AirParrot (App), 45
Animaker (App), 96
Annotation, 28-30, 42, 47, 94-95
Apple AirPlay (App), 45
Apple Messages (App), 7, 9
AppleTV (Hardware), 45
Archive tools, 49, 50-51, 55-58, 146-147
Arduino (Hardware), 121-122
Argumentation (written), 99, 157
Arlington National Cemetery, 139
Artificial intelligence, 111
Assembly (App), 94
Asynchronous group writing, 35
Augmented reality (AR), 63, 67, 79, 80-84, 171
Authenticity in learning activities, 47, 87
Automation, 112, 114
Awareness, 140
Backchannel, 142, 169
Bing (App), 145
Biteable (App), 96
Bookmarks (archive tools), 51-52, 56, 146
Bring Your Own Device (BYOD), i, 168, 172
Browsing effect, 55-58
Buffer (App), 52
Burns, Ken, 100
Camtasia (App), 43, 45
Capturecast (App), 43
Chat apps, 5-16, 20- 21, 36-37, 94-95, 141
Chat Fuel (App), 21
Chatbot, 13-21
Chromecast (Hardware), 45
Cloud (The). See Multi-platform
Cloud-based writing service, 33-34

CNN (Company), 14
Coaching, 115, 149
Coaching students, 115
Cognitive tutors, 18
Collaborative work, 35, 82, 162
Comments (Web), 36, 162
Communications tools. See Chat apps
Complexity, 74, 120
Computer-aided-design (CAD), 120
Conductive fabrics, 121
Conductive inks, 121
Content creation, 28, 57-58, 118, 134, 162
Contexts of technology use, 30, 36, 43, 47, 67, 80, 87-89, 137
Creative Commons (Company), 174-175
Creativity, 47, 74, 76, 77, 79, 120
Criteria for evaluating information on teh web, 155
Crowd collaboration. See Crowdsourcing
Crowdsourcing, 127-128, 164
Csikszentmihalyi, Mihaly, 76
Curation tools. See Archive tools
Customization, 49
Cyberbullying, 161-165
Data
 Privacy issues with data collection, 177
 Use in learning, 1, 7, 11, 24, 62, 65, 97, 98-107, 111-116, 141, 168, 175-177
Decision-making
 Computers making decisions, 113
Deliberation
 In online work, 128
Delicious (App), 52, 54, 56, 58, 146
Democratization of information, 129
Differentiation, 10, 101
Digital literacy, 8
Digital scanning apps, 170
Digital shelf life, 28
Digital storytelling, 24-30, 47, 87, 98-99, 106-107, 153, 170
Diigo (App), 52

Direct messages (DMs), 163
DMs. See Direct Messages
Easel.ly (App), 103
Edmodo (App), 2, 141, 162, 169, 176
Emboidment (psychology), 79, 81-82, 119
Engagement, 20, 28-30, 44, 76, 119, 172
Ephemerality, 8
Evaluation of information, 128, 149-159
Evernote (App), 52-54, 113
Facebook, 1, 7, 14-15, 21, 25-28, 49, 101, 107, 141, 146, 157, 162, 169, 176
    Facebook Developer Tools (Service), 21
    Facebook Live (App), 26
    Facebook Messenger (App), 7, 15
Facts and opinions on the web, 152
Fair use, 174
Feedback technologies, 158, 162
Fitbit (Hardware), 112-113
Flickr (App), 175
Free technologies, 173
Free technology, 175
Game, 7, 14-21, 29, 44, 61-89, 131, 135-141, 156, 176
Geographic education, 1, 7, 26, 31, 56-57, 65-69, 72, 81-82, 122, 129, 133, 146
Georgia Tech University, 20
GIFs, 25, 94-96
GIPHY (App), 95
Goodreads (App), 53
Google (App), 2, 7, 28, 33-41, 44, 46, 78, 80, 86, 89, 103, 141, 143-148, 154, 157, 171, 176
    Google Alerts (App), 147
    Google Books (App), 145
    Google Educators (App), 147
    Google News (App), 145
    Google Patents (App), 146
    Google Scholar (App), 145
    Google Allo (App), 7
    Google Cardboard (App), 86, 89
    Google Chart Tools (App), 103
    Google Docs (App), 2, 33-37, 44, 46, 141, 171
    Google Expeditions (App), 89
    Google Hangouts (App), 7

Griefing. See Cyberbullying
Group work, 10, 115
Hashtag (#), 27, 56, 146
HTC Vive (Hardware), 86
IFTTT (App), 11, 37, 52, 114-115
IKEA company, 81
Imgflip (App), 95
Implementation of technology, 142, 185
Infogr.am (App), 98, 103
Infographics, 101-108
Information literacy, 133
Information quality, 152, 154
Ingress (App), 69, 80
Instagram (App), 25, 28, 30, 94-96, 141, 169
Integration of technology, 37, 75, 98-99, 119, 121-122, 135, 141, 148
Interaction, 16, 18, 20, 30, 41, 46, 72, 81, 82, 87, 88, 138, 139, 141, 142, 164
Interactivity, 87, 112, 119, 170
Interface, 13, 16, 25, 40, 42, 45
Intersection of activities with convergent technologies, 136
Jill Watson (chatbot), 20
Joy (chatbot), 20
Khan Academy (App), 19
Kik (App), 6, 141, 169
Knewton (App), 19
Knowledge creation (participatory), 134
Layers of information, 47, 94
LEGO Mindstorms (App), 122
Libib (App), 53
Linux (App), 176
LittleBits (Hardware), 122
Live streaming, 26, 44
Liveblogging, 27
Loebner Prize, 17
M3D Printer (Company), 122
Machine interaction, 18, 20-21
Machine learning, 18, 21
Make Magazine, 117, 122
Maker Shed (Company), 122
Makerspace, 118, 121-123
Making & Maker Movement, 117, 119
Media choice, 8
MediaWiki, 176
Meme, 8, 25, 94-96

Merging group work, 36
Messaging apps. See Chat Apps
MetroMile (App), 112
Microblogging. See Short-snippet storytelling
Microsoft HoloLens (Hardware), 84
Microsoft Office 365 (App), 33
Minecraft (App), 71-80, 88
Mitsuku (App), 17
Mobile multimedia production, 8
Monitoring, 19, 20, 45, 139, 141-142, 163
Moodle (App), 176
Moovly (App), 96
Multimedia production, 170
Multimodal literacy. See Multiple literacies
Multi-platform, 9
Multiple literacies, 8
National Holocaust Museum (U.S.), 139
Natural language processing, 16
Netflix (App), 28
Niantic Inc. (Company), 62, 138
Nostalgia, 68
Notetaking technology, 52, 53, 172
Notifications, 37
Oculus Rift (Hardware), 86
On-demand, 10
Out-of-classroom learning opportunities, 76
Participation, 28, 30, 36, 131-133, 141, 163
  Barriers to, 30
Peer evaluation, 149
Personalization, 19
Photoshop Express (App), 93
Photoshop Sketch (App), 94
Physical movement and activity, 69
Physical prototyping, 118
PIC (Hardware), 121
Pinboard Technology, 53
Pinterest (App), 53-54
Pixelmator (App), 94
Pixlr (App), 42, 93, 103, 107
Place. See Geographic education
Plugins. See Add-ons
Pocket (App), 52, 54, 58
Pokémon Go (App), i, 61-69, 135-141
Poll Everywhere (App), 171
Polling. See Student feedback
Powtoon (App), 96
Prisma (App), 95, 97
Programmable microcontrollers & computers, 121
Publication, 58
QR code, 97-98, 170-171
Quality of information from the Internet, 151
Quora (App), 147
Raspberry Pi (Hardware), 121
Reddit (App), 147
Reflecting, 57, 58, 138
Reflector (App), 45, 46
Relevance of information on the web, 155
Repetitive tasks, 68
Robot, 15, 16, 20
Rules, 131
Samsung Gear VR (Hardware), 86
Screencapture, 40-47
Screencastify (App), 41, 43
Screencast-O-Matic (App), 42, 43
Screengrab. See Screencapture
Screenshot, 11, 42
ScreenStream Mirroring (App), 46
Search, 144
Sensory perception, 81, 87, 91, 119
Setting classroom expectations, 114
short-form storytelling. See Short-snippet storytelling
short-snippet storytelling, 23, 26, 30, 31
Short-snippet storytelling, 25, 31
Skype, 7
Smart technologies, 111
SMS, 7
Snagit (App), 43, 45, 46
Snapchat (App), 6, 8, 25-26, 28, 30, 94-96, 141, 169, 170
  Snapchat Discover (App), 28
  Snapchat Stories (App), 26
Snupps (App), 53
Social learning opportunities, 7, 25-28, 52, 57, 64, 68-69, 72, 74, 82, 88, 94-95, 133, 135, 137-141, 161-164, 167-170
Social networks, 25-26, 141, 162-163, 169
Socrative (App), 171
Sparkfun (Company), 122
Stack Exchange (App), 147

Stickers (images), 7, 94
Streak Trivia (App), 14-15, 18
Student feedback, 171
Tagging. See Tags
Tags, 52-55
Tags (archiving), 54-57
Teaching assistant as a chatbot, 20
Teaching strategies, 2
Technology goals, 2
Text-based, 17
The New York Times, 101
The Pokémon Company, 62, 69
Thinking pedagogically, 1, 2, 13, 73, 168
Trolling, 140, 161, 164
trolls. See Trolling
TweetChat (App), 146
Twitter (App), 25, 27-28, 30, 101, 107, 141, 146, 162, 169, 176
    Twitter Advanced Search (App), 146
Validity of information on the web, 156
Vessyl (Hardware), 112

ViewMaster (Hardware), 85, 87
Vimeo (App), 97
Vinci (App), 95
Virtual reality (VR), 80-89
Visual communication, 92
Visual demonstration, 41
Visualization, 39-40, 44-45, 83, 87-100, 103-104, 156
    Dimensions of, 105
Vloggers. See Vlogging
Vlogging, 28
VR goggles, 80, 84-86, 89
WhatsApp (App), 7, 141, 169
Wikipedia (App), 8, 16, 18, 25, 31, 56, 98, 121, 125-134, 149, 156, 176
    Community and participants, 131
    Talk page, 130, 132, 156
WordPress (App), 176
Yahoo (App), 145
YouTube (App), 25-30, 44, 67, 96-98, 122, 148, 152, 161, 170, 176
    YouTube Live (App), 26

About the Author

Jeremy Riel is an educational technologist with over a decade of experience in working with K-12 teachers and college faculty to implement educational technologies for teaching and learning. Additionally, he designs and researches digital systems for teacher professional development. Committed to the promises that online learning and digital literacy can bring, he has centered his work around the principled design of digital technologies for learning. Since 2012, he has served as the coordinator of the Faculty Assistance Center for Technology at the University of Illinois at Chicago College of Education, where he works with faculty and student teachers on educational technology projects. Find more out about his work and online writing at www.jeremyriel.com, or tweet him at @jeremyriel.

About The Educational Technology Network & The ETN Press

The Educational Technology Network (ETN) is a hub for resources and conversation around educational technology implementation. ETN contributors focus on the principled application of ed-tech in formal and informal learning environments and analyze the effects of today's latest tech trends. The ETN offers professional development programs, ongoing web series and podcasts, and a variety of support materials. Both veteran and student teachers alike can find useful, practical resources for using ed-tech smartly.

We encourage you to check out some of our professional development offerings, web videos, and other resources. Visit the ETN online at http://www.edtech.network and join the conversation!

The ETN Press publishes professional development and support resources for the ETN. Check out other print and e-book offerings by the ETN Press at http://press.edtech.network.

www.ingramcontent.com/pod-product-compliance
Lightning Source LLC
Chambersburg PA
CBHW071732080526
44588CB00013B/2004